THE MOST POWERFUL IDEA IN THE WORLD

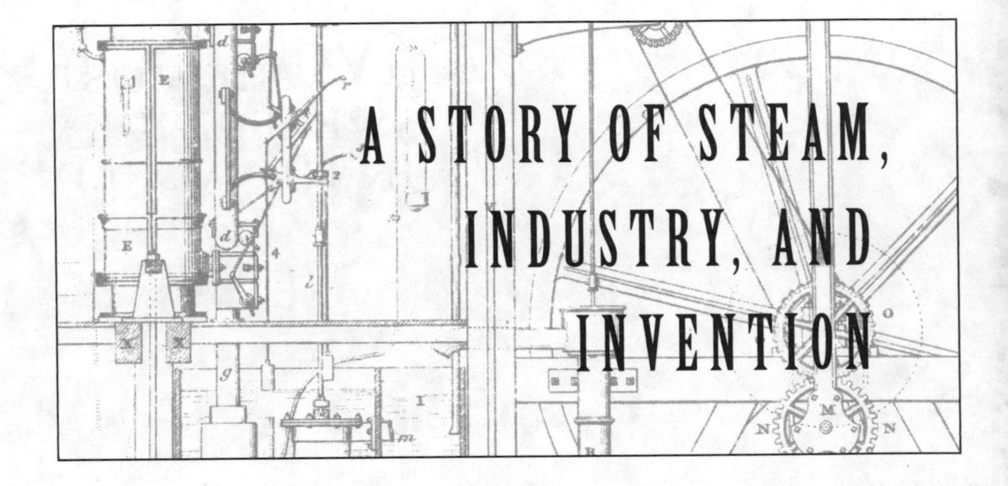

A STORY OF STEAM, INDUSTRY, AND INVENTION

The MOST POWERFUL IDEA *in the* WORLD

WILLIAM ROSEN

RANDOM HOUSE NEW YORK

Published in the United States by Random House,
an imprint of The Random House Publishing Group,
a division of Random House, Inc., New York.

RANDOM HOUSE and colophon are registered trademarks
of Random House, Inc.

LIBRARY OF CONGRESS CATALOGING-IN-PUBLICATION DATA

Rosen, William.
The most powerful idea in the world: a story of steam, industry, and invention /
William Rosen.
p. cm.
Includes bibliographical references and index.
ISBN 978-1-4000-6705-3
eBook ISBN 978-0-679-60361-0
1. Steam-engines—History. 2. Inventions—History. 3. Industrial revolution—
Great Britain—History. I. Title.
TJ46.R67 2010
909.81—dc22 2009041662

Printed in the United States of America on acid-free paper

www.atrandom.com

9 8 7 6 5 4 3 2 1

First Edition

Book design by Liz Cosgrove

To Quillan, Emma, and Alex—
my most valuable ideas

(and to Jeanine: my best one)

CONTENTS

LIST OF ILLUSTRATIONS

PROLOGUE

ROCKET

concerning ten thousand years, a hundred lineages, and two revolutions

ON THE GROUND FLOOR of the Science Museum in London's South Kensington neighborhood, on a low platform in the center of the gallery called "Making of the Modern World," is the most famous locomotive ever built.

Or what remains of it. *Rocket,* the black and sooty machine on display, designed and built in 1829 by the father and son engineers George and Robert Stephenson, no longer much resembles the machine that inaugurated the age of steam locomotion. Its return pipes are missing. The pistons attached to the two driving wheels are no longer at the original angle. The yellow paint that made it shine like the sun nearly two centuries ago is now not even a memory. Even so, the technology represented in the six-foot-long boiler, the linkages, the flanged wheels, and even in the track on which it rode are essentially the same as those it used in 1829. In fact, they are the same as those used for more than a century of railroading.

The importance of *Rocket* doesn't stop there. While the machine does, indeed, mark the inauguration of something pretty significant—two centuries of mass transportation—it also marks a culmination. Standing in front of *Rocket,* a museum visitor can, with a little imagination, see the thousand threads that lead from the locomotive back to the very beginning of the modern world. One such thread can be walked back to the first metalworkers who figured out how to cast the iron cylinders that drove *Rocket*'s wheels. Another leads to the discovery of the fuel that boiled the water inside that iron boiler. A third—the shortest, but probably the thickest—leads back to the discovery that boiling water could somehow be transformed into motion. One thread is, actually, thread: *Rocket* was built to transport cotton goods—the signature manufactured item of the first era of industrialization—from Manchester to Liverpool.

Most of the threads leading from *Rocket* are fairly straightforward, but one—the most interesting one—forms a knot: a puzzle. The puzzle of *Rocket* is why it was built to travel from Manchester to Liverpool, and not from Paris to Toulouse, or Mumbai to Benares, or Beijing to Hangzhou. Or, for that matter, since the world's first working model of a steam turbine was built in first-century Alexandria, why *Rocket* started making scheduled round trips at the beginning of the nineteenth century instead of the second.

Put more directly, why did this historical discontinuity called the Industrial Revolution—sometimes the "First" Industrial Revolution—occur when and where it did?*

The importance of that particular thread seems self-evident. At just around the time *Rocket* was being built, the world was ex-

* The term didn't really start to get traction until 1884, when a collection of lectures given by the economic historian Arnold Toynbee (the uncle of the famous one) at Balliol College starting in 1878 was posthumously published under the title *Lectures on the Industrial Revolution of the 18th Century in England, Popular Addresses, Notes, and Other Fragments.* This post hoc designation does have some arbitrariness to it; the most frequent textbook dates for the Industrial Revolution, 1760–1820, are a consequence of the fact that Toynbee's ostensible lecture subject was George III, whose regnal dates they are.

periencing not only a dramatic change *in* industry—what *The Oxford English Dictionary* calls "the rapid development in industry owing to the employment of machinery"—but also a transition *to* industry (or an industrial economy) from agriculture. Combining the two was not only revolutionary; it was unique.

"Revolutionary" and "unique" are both words shiny with overuse. Every century in human history is, in some sense, unique, and every year, somewhere in the world, something revolutionary seems to happen. But while love affairs, epidemics, art movements, and wars are all different, their effects almost always follow one familiar pattern or another. And no matter how transformative such events have been in the lives of individuals, families, or even nations, only twice in the last ten thousand years has something happened that truly transformed all of humanity.

The first occurred about 10,000 BCE and marks the discovery, by a global human population then numbering fewer than five million, that they could cultivate their own food. This was unarguably a world changer. Once humanity was tethered to the ground where its food grew, settled societies developed; and in them, hierarchies. The weakest members of those hierarchies depended on the goodwill of the strongest, who learned to operate the world's longest-lasting protection racket. Settlements became towns, towns became kingdoms, kingdoms became empires.

However, by any quantifiable measure, including life span, calories consumed, or child mortality, the lived experience of virtually all of humanity didn't change much for millennia after the Agricultural (sometimes known as the Neolithic) Revolution spread around the globe. Aztec peasants, Babylonian shepherds, Athenian stonemasons, and Carolingian merchants spoke different languages, wore different clothing, and prayed to different deities, but they all ate the same amount of food, lived the same number of years, traveled no farther—or faster—from their homes, and buried just as many of their children. Because while they made a lot more children—worldwide population grew a hundredfold between 5000 BCE and 1600 CE, from 5 to 500 million—they didn't make much of anything else. The best estimates for human productivity (a necessarily vague number) calculate annual per capita GDP, expressed in constant 1990 U.S. dollars, fluctuating between $400 and $550 *for seven thousand years.* The worldwide per capita GDP in 800 BCE—$543—is virtually identical to the number in 1600. The average person of William Shakespeare's time lived no better than his counterpart in Homer's.

The first person to explain why the average human living in the seventeenth century was as impoverished as his or her counterpart in the seventh was the English demographer Thomas Malthus, whose *Essay on the Principle of Population* demonstrated that throughout human history, population had always increased faster than the food supply. Seeking the credibility of a mathematical formula (this is a constant trope in the history of social science), he argued that population, unless unchecked by war, famine, epidemic disease, or similarly unappreciated bits of news, always increased geometrically, while the resources needed by that population, primarily food, always increased arithmetically.* The "Malthusian trap"—the term has been in general use for centuries—ensured that though mankind regularly discovered or invented more productive ways of feeding, clothing, transporting or (more frequently) conquering itself, the resulting population increase quickly consumed all of the surplus, leaving everyone in precisely the same place as before. Or frequently way behind, as populations exploded and then crashed when the food ran out. Lewis Carroll's Red Queen might have written humanity's entire history on the back of a matchbook: "Here, you see, it takes all the running you can do, to keep in the same place. If you want to get somewhere else, you must run at least twice as fast as that."

This is why *Rocket*'s moment in history is unique. That soot-blackened locomotive sits squarely at the deflection point where a line describing human productivity (and therefore human wel-

* "Geometric" and "arithmetic" are Malthus's terms; the modern equivalents are "exponential" and "linear."

fare) that had been flat as Kansas for a hundred centuries made a turn like the business end of a hockey stick. *Rocket* is when humanity finally learned how to run twice as fast.

It's still running today. If you examined the years since 1800 in twenty-year increments, and charted every way that human welfare can be expressed in numbers—not just annual per capita GDP, which climbed to more than $6,000 by 2000, but mortality at birth (in fact, mortality at any age); calories consumed; prevalence of infectious disease; average height of adults; percentage of lifetime spent disabled; percentage of population living in poverty; number of rooms per person; percentage of population enrolled in primary, secondary, and postsecondary education; illiteracy; and annual hours of leisure time—the chart will show every measure better at the end of the period than it was at the beginning. And the phenomenon isn't restricted to Europe and North America; the same improvements have occurred in every region of the world. A baby born in France in 1800 could expect to live thirty years twenty-five years less than a baby born in the Republic of the Congo in 2000. The nineteenth-century French infant would be at significantly greater risk of starvation, infectious disease, and violence, and even if he or she were to survive into adulthood, would be far less likely to learn how to read.

Think of it another way. A skilled laborer—a weaver, perhaps, or a blacksmith—in seventeenth-century England, France, or China spent roughly the same number of hours a week at his trade, producing about the same number of bolts of cloth, or nails, as his ten-times great-grandfather did during the time of Augustus. He earned the same number of coins a day and bought the same amount, and variety, of food. His wife, like her ten-times great-grandmother, prepared the food; she might have bought her bread from a village baker, but she made pretty much everything herself. She even made her family's clothing, which, allowing for the vagaries of weather and fashion, was largely indistinguishable from those of any family for the preceding ten centuries: homespun wool, with some linen if flax were locally available. The laborer and his wife would have perhaps eight or ten live offspring, with a reasonable chance that three might survive to adulthood. If the laborer chose to travel, he would do it on foot or, if he were exceptionally prosperous, by horse-drawn cart or coach, traveling three miles an hour if the former, or seven if the latter—again, the same as his ancestor—which meant that his world was not much larger than the five or six miles surrounding the place he was born.

And then, for the first time in history, things changed. And they changed at the most basic of levels. A skilled fourth-century weaver in the city of Constantinople might earn enough by working three hours to purchase a pound of bread; by 1800, it would cost a weaver working in Nottingham at least two. But by 1900, it took less than fifteen minutes to earn enough to buy the loaf; and by 2000, five minutes. It is a cliché, but nonetheless true, to recognize that a middle-class family living in a developed twenty-first-century country enjoys a life filled with luxuries that a king could barely afford two centuries ago.

This doesn't mean the transformation happened suddenly. A small but vocal minority of scholars doubts the reality of anything revolutionary, or even industrial, about the phenomenon. Recent studies have demonstrated far less growth in productivity and incomes during the period 1760–1820 than once thought, partly because the income of preindustrial Europe was a lot higher than previously believed. And indeed, Europe, from at least the ninth century onward, had urban centers, roads, and huge amounts of trade traveling along the latter to the former.

On the other hand, the fact that the transformation happened over the course of a century doesn't make it any less revolutionary. Clearly, *something* happened.

Not everyone believes that the something is the contraption sitting in that gallery in the Science Museum. There are, by popular consensus, more than two hundred different theories in general circulation purporting to explain the Industrial Revolution. They include the notion, first popularized by the pioneer sociologist Max Weber, that the Protestantism of Northern Europe was more congenial to innovation than Chinese Confucianism, or the

Catholicism of France and Southern Europe. Or that China's lack of access to raw materials, particularly coal, sabotaged an Asian Industrial Revolution. For those of a certain mindset, there is a theory that England's absence of internal tariffs and deficiency in landholding peasantry made the leap to industrialization a short one. Was industrialization the result of revenue from overseas colonies? Relatively high labor costs among the lower classes? Relatively large families among the upper classes? Class conflict? The *lack* of class conflict?

All of these explanations, even when reduced to bumper sticker size, are in some sense true. There are dozens of ways to untie a knot, and many will be referred to in later chapters of this book. Their only real liability, in fact, is that they tend to understate the most obvious explanation, which is that the Industrial Revolution was, first and foremost, a revolution in *invention*. And not simply a huge increase in the number of new inventions, large and small, but a radical transformation in the process of invention itself.

Given the importance of mechanical invention to every generation of humanity since some anonymous Sumerians stuck a pole through the center of a hollow tree trunk and rolled the first wheel past their neighbors, it's somewhat puzzling that it took so long to come up with a useful theory of just what invention *is*. Contemporary cognitive scientists have proposed a dozen different strategies and typologies of invention, but one of the most influential remains the eighty-year-old theory of an economic historian with the Dickensian name of Abbott Payson Usher.

Though dense, out of date, and little consulted today, *The History of Mechanical Inventions*, published by the then forty-six-year-old Usher in 1929, documents, at sometimes exhausting length, the ways in which humanity has engaged in a continuous process of improving life by inventing machines, from the earliest plows used by Middle Eastern farmers to the ships, engines, and railroads of the mid-nineteenth century (though, interestingly enough, not the age of electricity during which Usher wrote). Like *Origin of Species*, whose theory was buttressed by thousands of examples from the world of nature, *The History of Mechanical Inventions* contains an imposing list of examples, from the harnesses worn by prehistoric draft animals, Egyptian waterwheels and hand querns, to antique beam presses, medieval grain mills, water clocks, and, of course, the steam engine. But it does more than just chronicle human ingenuity. It also presents what is still the most analytically persuasive historical *theory* of invention: Usher, more than anyone else, gives us a toolkit that can be used to analyze and describe just how *Rocket* (and its component parts) was imagined, designed, and constructed.

Before Usher, historians of science hadn't wandered very far from the same two paths that general historians had trod before them. The first is popularly known as the "Great Man" theory of history, in which events are understood through the actions of a few major actors—in this context, the "Great Inventor" theory—while the second perceives those same events as consequences of immutable laws of history; for the history of science and technology, this frequently meant explaining things as a sort of evolution of inventions by natural selection. Usher hated them both. He was, philosophically and temperamentally, a small-*d* democrat who was utterly convinced that the ability to invent was widely distributed among ordinary people, and that the impulse to invent was everywhere.

If the phenomenon of invention were as natural as breathing, one might expect that it would—like breathing—behave pretty much the same whether it occurred in second-century Egypt or eighteenth-century England, and so indeed it did for Usher. To him, every invention inevitably followed a four-step sequence:

1. Awareness of an unfulfilled need;
2. Recognition of something contradictory or absent in existing attempts to meet the need, which Usher called an "incomplete pattern";
3. An all-at-once insight about that pattern; and
4. A process of "critical revision" during which the insight is tested, refined, and perfected.

Usher is an invaluable guide to the world of inventing, and in the pages that follow, his step-by-step description of the inventive process will be referred to many times. But precisely because his sequence applies to everything from Neolithic digging sticks to automated looms, it cannot explain *why*—in the unforgettable line of the imagined schoolboy introducing T. S. Ashton's short but indispensable history of the Industrial Revolution—"About 1760, a wave of gadgets swept over England." If the process of thinking up "gadgets" was, at bottom, the same for Archimedes, Leonardo, and James Watt, why did it take until the middle of the eighteenth century for a trickle to become a wave?

Even defining the Industrial Revolution as a wave of gadgets doesn't, by itself, place steam power—*Rocket*'s motive force—at the crest of that wave. After all, the early decades of European industrialization were largely driven by water and wind rather than steam. As late as 1800, Britain's water mills were producing more than three times as much power as its steam engines, and this book could, conceivably, have begun not with *Rocket*, but with another display in the "Making of the Modern World" gallery: Richard Arkwright's cotton spinning machine, known as the "water frame" because of its power source.* Nonetheless, the steam engine *was* the signature gadget of the Industrial Revolution, though not because it represented a form of power not dependent on muscle; both waterwheels and windmills had already done that. Nor was it the steam engine's enormous capacity for rapid improvement—far greater than either water or wind power.

The real reason steam power dominates every history of the Industrial Revolution is its central position connecting the era's technological and economic innovations: the hub through which the spokes of coal, iron, and cotton were linked. The steam engine was first invented to drain the mines that produced the coal burned in the engine itself. Iron foundries were built to supply the boilers for the steam engines that operated forges and blast furnaces. Cotton traveled to the British Isles on steamships, was

* For more about Arkwright—*much* more, in fact—see chapter 10.

spun into cloth by steam-powered mills, and was brought to market by steam locomotives. Thousands of innovations were necessary to create steam power, and thousands more were utterly dependent upon it, from textile factories—soon enough, even the water frame was steam-driven—to oceangoing ships to railroads. After thousands of years of searching for a perpetual motion machine, the inventors of the steam engine at *Rocket*'s heart created something even better: a perpetual *innovation* machine, in which each new invention sparked the creation of a newer one, ad—so far, anyway—infinitum.

Perpetual technological innovation is so much a part of contemporary life that it is difficult even to imagine the world without it. It is the modern world, however, that is historically anomalous. Hundreds of different cultures had experienced bursts of inventiveness and economic growth before the eighteenth century—bursts they were unable to sustain for more than a century or so. Imagine, for example, how different the last eight hundred years might have been had the Islamic Golden Age—whose inventors were responsible for everything from crankshaft-driven windmills and water turbines to the world's most advanced mechanical clocks—survived the thirteenth century. Instead, like all the world's earlier explosions of invention, it, in the words of one of the phenomenon's most acute observers, "fizzled out." One unique characteristic of the eighteenth-century miracle was that it was the first that didn't.

The other one, and the real reason that the threads leading from *Rocket* form such a challenging knot, is that the miracle was, overwhelmingly, produced by English-speaking people. *Rocket* incorporates hundreds of inventions, small and large—safety valves, feedback controls, return flues, condensers—to say nothing of the iron foundries and coal mines that supplied its raw materials. If one could magically edit out those steam engines invented in Italy, or Sweden, or—more important—France, or China, *Rocket* would still run. If the same magic were applied to those invented in England, Scotland, Wales, and America, the platform in the Science Museum would be empty.

That is a puzzle for which there is no shortage of proposed solutions (see Industrial Revolution, Theories of, above). The one proposed by the book you hold in your hands can be boiled down to this: *The best explanation for the preeminence of English speakers in lifting humanity out of its ten-thousand-year-long Malthusian trap is that the Anglophone world democratized the nature of invention.*

Even simpler: Before the eighteenth century, inventions were either created by those wealthy enough to do so as a leisure activity (or to patronize artisans to do so on their behalf), or they were kept secret for as long as possible. In England, a unique combination of law and circumstance gave artisans the incentive to invent, and in return obliged them to share the knowledge of their inventions. Virginia Woolf's famous observation—that "on or about December, 1910, human character changed"—was not only cryptic, but about a century off. Or maybe two. Human character (or at least behavior) was changed, and changed forever, by seventeenth-century Britain's insistence that ideas were a kind of property. This notion is as consequential as any idea in history. For while the laws of nature place severe limits on the total amount of gold, or land, or any other traditional form of property, there are (as it turned out) no constraints at all on the number of potentially valuable ideas. The result was that an entire nation's unpropertied populace was given an incentive to produce them, and to acquire the right to exploit them.

OBSERVE ANY GROUP OF people, and you can, if you're so inclined, find clues to their ancestry in their hair or skin color. Examine blood or skin cells under a microscope, and you can learn still more; sequence your subjects' DNA, and you'll know quite a bit indeed, including the portion of the planet where their many-times great-grandparents lived, and genetic relationships between and among them.

Stand in front of *Rocket,* and you'll likely see "only" a rather complicated machine. But examine it with a historian's microscope, and it will become clear that the "genetic sequence"* of the locomotive, and of the Industrial Revolution it exemplifies, comprises a hundred lineages taken from a dozen different disciplines, as ornate and as complicated as the family tree of a European royal family. The birth of steam depended on a new understanding of the nature of air, and its absence; on an empirical, not yet scientific, understanding of thermodynamics; and on a new language of mechanics describing how matter moves other matter. It was utterly dependent on a new "iron age" inaugurated by several generations of a single English family; a change in the understanding of national wealth, itself a contribution from the Scottish Enlightenment, and of the special character of water as a medium for storing and releasing heat. Perhaps the most important father of the steam engine was the notion that ideas were property, itself the progeny of one of England's greatest jurists, and her most famous political philosopher. The threads tied to *Rocket* lead back to an Oxford college and a Birmingham factory, to Shropshire forges and Cornish mines, to a Yorkshire monastery and a Virginia flour mill, to a Westminster courtroom and a Piccadilly locksmith. Those threads end at some of history's great eureka moments: an Edinburgh professor's discovery of carbon dioxide; an expatriate American's demonstration that heat and motion are two ways of thinking about the same thing; even a Greek fisherman's discovery of a first-century calculating machine. All of them—metallurgy and legal advocacy, chemistry and kinematics, physics and economics—are on display in the pages that follow.

But most of these pages are about invention itself. No one can stand in front of *Rocket* for long without pondering the history of this peculiarly human activity, its psychology, economics, and social context. The narrative of steam may be constrained by the limits of mechanics, but it is defined by the behavior of inventors, and the pages that follow attempt to explore not only what inventors actually *do,* but what happens inside their skulls while they

* The term is a favorite of A. P. Usher.

do it, touching on recent discoveries in neurobiology, cognitive science, and evolutionary sociology.

Ever since humanity became bipedal, it has invented things. Stone tools in east Africa 2.4 million years ago, pottery in Anatolia eight thousand years ago. Five thousand years later, Archytas of Tarentum described the pulley, and Archimedes—probably—invented the lever, screw, and wedge. For a thousand centuries, the equation that represented humanity's rate of invention could be plotted on an X-Y graph as a pretty straight line; sometimes a little steeper, sometimes flat. Then, during a few decades of the eighteenth and nineteenth centuries, in an island nation with no special geographic resource, a single variable changed in that equation. The result was a machine that changed everything, up to and including the idea of invention itself. The components of *Rocket,* and therefore the Industrial Revolution, are not gears, levers, and boilers, but *ideas* about gears, levers, and boilers—the most important ideas since the discovery of agriculture.

But here is the difference: Many societies discovered agriculture independently, from the Fertile Crescent to the Yangtze to the Indus River Valley. The miracle of sustainable innovation has a single source, a single time and place where mankind first made the connection between invention, power, and wealth, and discovered the most powerful idea in the world.

THE MOST POWERFUL IDEA IN THE WORLD

CHAPTER ONE

HANGES IN MOSPHERE

andria failed to inspire, and how a
l; the spectacle of two German hemi-
n horses; and the critical importance

gham, you take the M5 south
nd then change to the A417,
hwest, then southeast, for an-
no apparent reason, into the
urbage, you turn left at the
her mile, across the railroad
on for making this three-hour
ng turns) is visible for the last
ildings next to a sixty-foot-tall

Wiltshire contains the oldest
g the job for which it was de-
operated beam pumps twelve
ght-foot-high locks along the
von Canal. The engine itself,
s still visible on the engine

beam—is so called because it was the second engine with a forty-two-inch cylinder produced by the Birmingham manufacturer Boulton & Watt. It was entered in the company's order book on January 11, 1810, and installed almost precisely two years later. Except for a brief time in the 1960s, it has run continuously ever since.

First encounters with steam power are usually unexpected, inadvertent, and explosive; the cap flying off a defective teakettle, for example. No surprise there; the expansive property of water when heated past a certain point was known for thousands of years before that point was ever measured, and to this day it's what drives the turbine that generates most of our electricity, including that used to power the light by which you are reading this book. The relationship between the steam power of a modern turbine and the kind used to pump the water out of the Kennet and Avon Canal is, however, anything but direct. By comparison, the mechanism of engine 42B is a thing of Rube Goldberg–like complexity, with levers, cylinders, and pistons yoked together by a dozen different linkages, connecting rods, gears, cranks, and cams, all of them moving in a terrifyingly complicated dance that is at once fascinating, and eerily quiet—enough to occupy the mechanically inclined visitor, literally, for hours. When the engine is "in steam," it somehow causes the twenty-six-foot-long cast iron beams to move, in the words of Charles Dickens, "monotonously up and down, like the head of an elephant in melancholy madness."

There is, however, something odd about the beams, or rather about the pistons to which they are attached. The pistons aren't just being driven *up* by the steam below them. The power stroke is also *down:* toward the steam chamber. Something is sucking the pistons downward. Or, more accurately, nothing is: a vacuum.

Using steam to create vacuum was not the sort of insight that came an instant after watching a teakettle lid go flying. It depended, instead, on a journey of discovery and diffusion that took more than sixteen centuries. By all accounts the trip began sometime in the first century CE, on the west side of the Nile Delta, in the Egyptian city of Alexandria, at the Mouseion, the great university at which first Euclid and then Archimedes studied, and where, sometime around 60 CE, another great mathematician lived and worked, one whose name is virtually always the first associated with the steam engine: Heron of Alexandria.

The *Encyclopaedia Britannica* entry for Heron—occasionally, Hero—is somewhat scant on birth and death dates; as is often the case with figures from an age less concerned with such trivia, it uses the abbreviation "fl." for the latin *floruit,* or "flourished." And flourish he did. Heron's text on geometry, written sometime in the first century but not rediscovered until the end of the nineteenth, is known as the *Metrika,* and includes both the formula for calculating the area of a triangle and a method for extracting square roots. He was even better known as the inventor of a hydraulic fountain, a puppet theater using automata, a wind-powered organ, and, most relevantly for engine 42B, the *aeolipile,* a reaction engine that consisted of a hollow sphere with two elbow-shaped tubes attached on opposite ends, mounted on an axle connected to a tube suspended over a cauldron of water. As the water boiled, steam rose through the pipe into the sphere and escaped through the tubes, causing the sphere to rotate.

Throughout most of human history, successful inventors, unless wealthy enough to retain their amateur status, have depended on patronage, which they secured either by entertaining their betters or glorifying them (sometimes both). Heron was firmly in the first camp, and by all accounts, the aeolipile was regarded as a wonder by the wealthier classes of Alexandria, which was then one of the richest and most sophisticated cities in the world. Despite the importance it is given in some scientific histories, though, its real impact was nil. No other steam engines were inspired by it, and its significance is therefore a reminder of how quickly inventions can vanish when they are produced for a society's toy department.

In fact, because the aeolipile depended only upon the expansive force of steam, it should probably be remembered as the first in a line of engineering dead ends. But if the inspirational value of

Heron's steam turbine was less than generally realized, that of his writings was incomparably greater. He wrote at least seven complete books, including *Metrika,* collecting his innovations in geometry, and *Automata,* which described a number of self-regulating machines, including an ingenious mechanical door opener. Most significant of all was *Pneumatika,* less for its descriptions of the inventions of this remarkable man (in addition to the aeolipile, the book included "Temple Doors Opened by Fire on an Altar," "A Fountain Which Trickles by the Action of the Sun's Rays," and "A Trumpet, in the Hands of an Automaton, Sounded by Compressed Air," a catalog that reinforces the picture of Heron as antiquity's best toymaker) than for a single insight: that the phenomenon observed when sucking the air out of a chamber is nothing more than the pressure of the air around that chamber. It was a revelation that turned out to be utterly critical in the creation of the world's first steam engines, and therefore of the Industrial Revolution that those engines powered.

The idea wasn't, of course, completely original to Heron; the idea that air is a source of energy is immeasurably older than science, or even technology. Ctesibos, an inventor and engineer born in Alexandria three centuries before Heron, supposedly used compressed air to operate his "water organ" that used water as a piston to force air through different tubes, making music.

Just as the ancients realized that moving air exerts pressure, they also recognized that its absence did something similar. The realization that sucking air out of a closed chamber creates a vacuum seems fairly obvious to any child who has ever placed a finger on top of a straw—as indeed it was to Heron. In the preface to *Pneumatika,* he wrote,

> if a light vessel with a narrow mouth be taken and applied to the lips, and the air be sucked out and discharged, the vessel will be suspended from the lips, the vacuum drawing the flesh towards it that the exhausted space may he filled. It is manifest from this that there was a continuous vacuum in the vessel. . . .

thus producing what a modern scholar has called a "very satisfactory theory of elastic fluids."

Satisfactory to a twenty-first-century child, and a first-century mathematician, but not, unfortunately, for a whole lot of people in between. To them, the idea that space could exist absent any occupants, which seems self-evident, was evidently not, and the reason was the dead hand of the philosopher-scientist who tutored Alexandria's founder. Aristotle argued against the existence of a vacuum with unerring, though curiously inelegant, logic. His primary argument ran something like this:

1. If empty space can be measured, then it must have dimension.
2. If it has dimension, then it must be a body (this is something of a tautology: by Aristotelian definition, bodies are things that have dimension).
3. Therefore, anything moving into such a previously empty space would be occupying the same space simultaneously, and two bodies cannot do so.

More persuasive was the argument that a void is "unnecessary," that since the fundamental character of an object consists of those measurable dimensions, then a void with the same dimensions as the cup, or horse, or ship occupying it is no different from the object. One, therefore, is redundant, and since the object cannot be superfluous, the void must be.

It takes millennia to recover from that sort of unassailable logic, temptingly similar to that used in *Monty Python and the Holy Grail* to demonstrate that if a woman weighs as much as a duck, she is a witch. Aristotle's blind spot regarding the existence of a void would be inherited by a hundred generations of his adherents. Those who read the work of Heron did so through an Aristotelian scrim on which was printed, in metaphorical letters twenty feet high: NATURE ABHORS A VACUUM.

Given that, it is something of a small miracle that *Pneumatika,* and its description of vacuum, survived at all. But survive it did, like so many of the great works of antiquity, in an Arabic translation, until around the thirteenth century, when it first appeared in Latin. And it was another three hundred years until a really influential translation arrived, an Italian edition translated by Giovanni Batista Aleotti d'Argenta and published in 1589. Aleotti's work, and subsequent translations of his translation into German, English, and French (plus five more in Italian alone), demonstrate both the demand for and availability of the book. Aleotti, an architect and engineer, was practical enough; in his annotations to his translation of the *Pneumatika,* he mentions the difficulty of removing a ramrod from a cannon with its touchhole covered because of the pressure of air against the vacuum therefore created—a phenomenon that could only exist if air were compressible and vacuum possible. It is testimony to the weight of formal logic that even with the evidence in front of his nose, Aleotti was still intellectually unable to deny his Aristotle.

If Aleotti was unaware of the implications of Heron's observations, he was indefatigable in promoting them, and by the seventeenth century, it can, with a wink, be said that *Pneumatika* was very much in the air, in large part because of the Renaissance enthusiasm for duplicating natural phenomena by mechanical means, the era's reflexive admiration for the achievements of Greek antiquity. The scientist and philosopher Blaise Pascal (who modeled his calculator, the Pascaline, on an invention of Heron's) mentioned it in *D'esprit géometrique,* as did the Oxford scholar Robert Burton in his masterpiece, *Anatomy of Melancholy:* "What is so intricate, and pleasing as to peruse . . . Hero Alexandrinus' work on the air engine." But nowhere was Aleotti's translation more popular than the city-state of Firenze, or Florence.

Florence, in the year 1641, had been essentially the private fief of the Medici family for two centuries. The city, ground zero for both the Renaissance and the Scientific Revolution, was also where Galileo Galilei had chosen to live out the sentence imposed by the Inquisition for his heretical writings that argued that the earth revolved around the sun. Galileo was seventy years old and living in a villa in Arcetri, in the hills above the city, when he read a book on the physics of movement titled *De motu* (sometimes *Trattato del Moto*) and summoned its author, Evangelista Torricelli, a mathematician then living in Rome. Torricelli, whose admiration for Galileo was practically without limit, decamped in time not only to spend the last three months of the great man's life at his side, but to succeed him as professor of mathematics at the Florentine Academy. There he would make a number of important contributions to both the calculus and fluid mechanics. In 1643, he discovered a core truth in the behavior of liquids in motion, known as Torricelli's theorem, that is still used to calculate the speed of a fluid when it exits the vessel that contains it. He made fundamental contributions to the development of the calculus, and to the geometry of the cycloid (the path described by a point on a rolling wheel). Less typically, he embarked on a series of investigations whose results were, literally, revolutionary.

In those investigations, Torricelli used a tool even more powerful than his well-cultivated talent for mathematical logic: He did experiments. At the behest of one of his patrons, the Grand Duke of Tuscany, whose engineers were unable to build a sufficiently powerful pump, Torricelli designed a series of apparatuses to test the limits of the action of contemporary water pumps. In spring of 1644, Torricelli filled a narrow, four-foot-long glass tube with mercury—a far heavier fluid than water—inverted it in a basin of mercury, sealing the tube's top, and documented that while the mercury did not pour out, it did leave a space at the closed top of the tube. He reasoned that since nothing could have slipped past the mercury in the tube, what occupied the top of the tube must, therefore, be nothing: a vacuum.

Even more brilliantly, Torricelli reasoned, and then demonstrated, that the amount of space at the top of the tube varied at different times of the day and month. The only explanation that accounted for his observations was that the variance was caused by the pressure of air; the more pressure on the open reservoir of mercury at the base of the tube, the higher the mercury rose within. Torricelli had not only invented, more or less accidentally, the first barometer; he had demonstrated the existence of air pressure, writing to his colleague Michelangelo Ricci, "I have already called attention to certain philosophical experiments that are in progress . . . relating to vacuum, designed not just to make a vacuum but to make an instrument which will exhibit changes in the atmosphere . . . we live submerged at the bottom of an ocean of air. . . ."

Torricelli was not, even by the standards of his day, a terribly ambitious inventor. When faced with hostility from religious authorities and other traditionalists who believed, correctly, that his discovery was a direct shot at the Aristotelian world, he happily returned to his beloved cycloids, the latest traveler to find himself on the wrong side of the boundary line between science and technology.

But by then it no longer mattered if Torricelli was willing to leave the messiness of physics for the perfection of mathematics; vacuum would keep mercury in the bottle, but the genie was already out. Nature might have found vacuum repugnant for two thousand years, but Europe was about to embrace it.

ON NOVEMBER 20, 1602, in Magdeburg, a town in Lower Saxony, hard by the Elbe River, the former Anna von Zweidorff, by then the wife of a prosperous landowner named Hans Gericke, gave birth to a son, Otto. This was something like being born in Mogadishu, Somalia, in 1975: When Otto was sixteen years old, the armies of the last great religious war in European history began marching and countermarching across Germany, enforcing orthodoxy at the end of a pike in what became known as the Thirty Years War. Magdeburg, which had been a bastion of Protestantism ever since Martin Luther had visited in 1524, became a target for the armies of the Catholic League, not once, but half a dozen times; in 1631, the troops of Count Johann Tilly sacked the city, killing more than twenty thousand. By the time the various treaties that comprised the Peace of Westphalia were signed in 1648, the city was home to fewer than five hundred war-weary survivors. One of them was Otto Gericke, home from his studies in Leipzig, Jena, and Leiden, now a military engineer who was enlisted to help rebuild the city, and had been named one of its four mayors. He was, entirely as one might expect, eager to turn his talents to more peaceful pursuits.

Though evidently unaware of the details of Torricelli's experiments, he was headed down the same path, intending to demonstrate the power of a vacuum and therefore the weight of air. By 1650 or so, he had built the *Magdeburger windbüchse,* which looked like a gun but worked like a vacuum pump, a piston encased in a cylinder with an ingenious one-way flap valve that kept the cylinder airtight once the piston was withdrawn and was rightly regarded as one of the "technical wonders of its time." It was, however, barely an appetizer for what came next. For in 1652, Gericke, fascinated by the elasticity and compressibility of air, was to produce some of the most famous experimental apparatuses in history.

The original copper objects that came to be known as the Magdeburg hemispheres are on view at the Deutsches Museum in Munich, looking today a bit like oversized and battered World War I army helmets, with a dark bronze patina caused by nearly four hundred years of oxidation. Ropes dangle from half a dozen iron fasteners on both, and one holds a tube designed to mate with Gericke's vacuum pump. When Gericke constructed them in 1665, the ropes were tied to the harnesses of a team of horses, and the copper shone like a mirror. The reasons had more to do with theater than science. With the smooth rims of the hemispheres coated with grease, the air pumped out of the globe, and the horses urged in opposite directions, the show was irresistible. Its first appearance was in 1654, in front of the Imperial Diet in Regensburg, where Gericke tied his ropes to thirty horses—fifteen attached to either hemisphere—and demonstrated their inability to pull the pieces apart. That was followed by similar entertainments in 1656 in Magdeburg (with sixteen horses), in 1657 before the emperor's court in Vienna, and most famously of all, in 1664, before the German elector Friedrich Wilhelm, who was amazed to see twenty-four horses straining to pull apart a twenty-inch globe held together only by air pressure.*

The Magdeburg hemispheres are deservedly some of the most famous experimental devices of all time, and versions are still used in science classrooms to this day. But their fame owes at least as much to showmanship as to any intrinsic contribution to the physics of vacuum. In 1661, Gericke performed a far more sophisticated, though less well remembered, experiment. It consisted of two suspended platforms connected by a single rope, each under a pulley, with both pulleys suspended from a horizontal beam. On one he placed an airtight chamber with a close-fitting piston; on the other, a measured amount of lead weight. As the air was pumped out of the chamber, the piston was forced down by the weight of atmosphere, and the weight raised by the same amount—the first practical application of the power of the vacuum, well recounted in his 1672 book, *Experimenta nova, ut vocantur, Magdeburgica de vacuo spatio.*

But it was the hemispheres that, in the end, mattered. They are the reason Emperor Leopold I knighted Gericke in 1666, making him Otto von Guericke (including the unexplained introduction of the *u* to his name). It was the hemispheres that a German Jesuit and mathematician named Gaspar Schott saw at the 1654 demonstration, and that initiated an admiring correspondence between Schott and Gericke. And it was the hemispheres that were featured in Schott's 1657 book, with the intimidating title *Mechanicahydraulica-pneumatica,* which contained a description of both the vacuum pump and the hemispheres (and included a drawing eerily similar to the logo used by Levi Strauss to testify to the inability of even whipped horses to pull a pair of jeans apart). And of course the hemispheres mark another fork in the road for the idea powering engine 42B on its way from conti-

* Part of the story of the Magdeburg hemispheres remains a bit of a mystery. Even if Gericke had been able to achieve a perfect vacuum—unlikely, with the equipment he had at hand—the total air pressure at sea level on a globe twenty inches in diameter would be a bit less than five thousand pounds—a lot, but not too much for thirty horses.

nental Europe to Britain, where Schott's book traveled almost as soon as it was published.

ENGLAND, IN THE MIDDLE of the seventeenth century, had not witnessed the brutal devastation that had been visited upon Gericke's homeland by the Thirty Years War, but it had not exactly been a model of peaceful coexistence either. A dispute between King and Parliament over their respective degrees of authority exploded into civil war in 1643; it had been temporarily suspended by the execution of Charles I and the exile of his son, Charles II, but not before a hundred thousand men, women, and children were dead. One of the Civil War's less dramatic but equally far-reaching consequences was that the various colleges at Oxford, which had been the king's base of operations for much of the war, had walked a delicate line between their traditional and reflexive support for the monarchy and prudent obedience to its replacements: first the Commonwealth, and then the de facto dictatorship of Oliver Cromwell. By the time England, and Oxford, had received copies of Schott's book, they had been without a king for years, and the town's scholars, two in particular, were more interested in persuading nature to give up her secrets than in forcing their countrymen to choose a sovereign.

It seems almost indecently apt that Robert Hooke and Robert Boyle were among the first, and certainly the most important, Englishmen to learn of Gericke's experiments. The aptness is not due entirely to their interest in vacuum; these wildly inventive, almost ridiculously prolific men were interested in practically everything. A brief list of their respective achievements would include the discovery of the Law of Elasticity; the founding of the science of experimental chemistry; the invention of the microscope; the discovery of the basic law governing the behavior of gases; the first observation of the rotation of both Jupiter and Mars; the discovery of the inverse-square law of gravity; the authorship of some of the seventeenth century's most profound Christian apologetics; and the founding of the world's first scientific society.

Their link began in 1659 or so, when Boyle, a brilliant and wealthy aristocrat, hired Hooke, a brilliant and impecunious scholarship student, to improve on Gericke's vacuum pump. The improvement that Boyle had in mind was critical: He needed a machine that would not merely demonstrate the existence of a vacuum for the entertainment of European aristocrats, but would allow him to investigate its characteristics. Hooke's answer was the *machine Boyleana,* an experimental device that would reveal what was happening inside the vacuum chamber and allow manipulation of it. Boyle had earlier hired the now forgotten Ralph Greatorex ("the leading pumping engineer in England") to achieve these goals, but where he had failed, Hooke succeeded. His design incorporated a glass vessel and two cone-shaped brass stoppers that, when coated with oil, could be rotated, pulling a thread that could be attached to the clapper of a bell, the wick of a candle—to anything, in short, that might be part of a viable experiment on the nature of vacuum.

All by itself, Hooke and Boyle's series of vacuum experiments, described in the 1660 publication of *New Experiments Physico-Mechanical, Touching the Spring of the Air and Its Effects,* would have bought them an entry in the history of steam power. In their hands, the machine Boyleana made basic discoveries into the properties of sound—when air was removed from the chamber, so too was the sound of a bell within it—of animal respiration, and of combustion. The experiments conducted by the two men produced the law of physics that still bears Boyle's name,* and the demonstration that the volume of a gas at constant temperature is inversely proportional to pressure (with the corollary that increasing temperature equals increased pressure) is an insight of some significance for the road leading from Torricelli's mercury tube to engine 42B.

However, the most significant characteristic of the two men's work—the one that best reveals why the road to steam power was

* Though it should be noted that in 1676, the French physicist and priest Edmé Mariotte independently discovered "Boyle's" law, and that in many European countries, the same equation is known as Mariotte's Law.

thereafter almost entirely an English one—is the fact that Boyle hired Hooke.

BEGIN WITH THEIR BEGINNINGS: Robert Boyle was one of the younger sons of an earl, born in Lismore Castle and educated at Eton, in Switzerland, and in France. By the time he returned from Florence in 1642 (where he read Galileo's *Dialogue on the Two Chief World Systems* and began a lifelong devotion to mechanical explanations: in his words "those two grand and most catholic principles, matter and motion"), his father had died, leaving him a Dorsetshire manor and sufficient income from his Irish estates to study whatever part of "matter and motion" took his fancy. Hooke was born to a modest curate on the Isle of Wight, who left him just enough to purchase an apprenticeship with a portrait painter. Boyle arrived in Oxford in 1654 as a gentleman scholar; Hooke made his way to Oxford a year later, a scholarship student eager for anything to supplement his very modest stipend.

The two did share an affinity for the royalist cause, though not especially for the High Anglicanism associated with it. Boyle, in particular, was a devoted Protestant, well remembered for his piety, who famously argued (in *The Christian Virtuoso*) that devoutness did not forbid study of natural phenomena, but rather demanded it. His advocacy of experiment and experience—in brief, empiricism—as the best method for explaining the world was partly a response to the materialism (halfway to atheism, in the view of Boyle's Oxford colleague, Seth Ward) of Thomas Hobbes, who returned the favor, sneering at Boyle's work, which he called "engine philosophy."

Robert Hooke's philosophy, on the other hand, seems to have been driven more by a need for recognition than salvation. For all his extraordinary range of achievements (not only was he Christopher Wren's surveyor and colleague during the rebuilding of London after the Great Fire of 1666, an early advocate of evolutionary theory, the first to see that organic matter was made up of the building blocks that he named "cells," and probably England's most gifted mathematician, able to turn his hand to everything from describing the catenary curve of the ideal arch to the best way to trim sails), he is frequently remembered today, as he was known during his lifetime, as the world's best second fiddle. The shadow cast by Wren, by Boyle, and even by Isaac Newton, with whom Hooke engaged in a long-running and ultimately futile dispute over the authorship of the law of gravitational attraction, is unaccountable without considering the class difference between them. James Aubrey, the seventeenth-century memoirist, paid Hooke something of a backhanded compliment when he called him "the best Mechanick this day in the world."

When the informal assembly at Oxford whose meetings were generally led by the clergyman John Wilkins was chartered, two years after the Restoration of Charles II in 1660, as the Royal Society of London for the Improvement of Natural Knowledge, each Fellow was explicitly to be a "*Gentleman,* free, and unconfin'd." Hooke's need to make a living disqualified him from fellowship, though his talent made him indispensable. The solution—he was appointed to the salaried position of curator of experiments for the Royal Society in 1662—made him the first scientist in British history to receive a salary, though the salary in question was long in coming. It took until 1665, when Hooke was appointed professor of geometry at Gresham College at an annual stipend of £50 for life; the Royal Society then coughed up another £30, to make good on their original promise to Hooke of £80 a year.

Robert Hooke's pioneer status makes him a persuasive bridge between technology and science, which was in 1665—and for decades thereafter, in Britain and everywhere else—still the province of amateurs. Hooke spent his life in an occasionally successful search for both recognition and recompense, attempting, among other things, to turn his Law of Elasticity into ownership of the watch escapement, whose spring-loaded movement was a direct outgrowth of the Law.* When he died, his frugally ap-

* Tellingly, in order to keep his discovery secret, and so secure his status as its discoverer, he first published the Law in the form of an anagram.

pointed apartments contained a considerable amount of cash, largely earned from his surveying, contributing to a probably false reputation as a bit of a miser, but his attitude toward invention seems to be, in its way, as significant an innovation as his vacuum pump.

While Boyle is traditionally remembered as the more important transitional figure in the development of steam power, he exhibited a strong prejudice in favor of those whose experiments were entirely in service of the search for truth, as opposed to those "mere Empiricks" and "vulgar chemists" simply trying to "produce effects." This distinction makes his position clear in the never-ending debate between pure and applied science—really, between science and invention—that was already thousands of years old by Boyle's day.

The debate continues into our own day. Which is why it is Robert Hooke's life, rather than Boyle's, that leads from Torricelli (whose promising start on the potential uses of vacuum were forestalled by conservative Aristotelianism) and von Guericke (whose undoubted talent for innovation is mostly remembered as a circus act) on the path to engine 42B, and to *Rocket*.

The next steps on that path would take the technology of steam and vacuum irrevocably into the world of commerce.

CHAPTER TWO

A GREAT COMPANY OF MEN

concerning the many uses of a piston; how the world's first scientific society was founded at a college with no students; and the inspirational value of armories, Nonconformist preachers, incomplete patterns, and snifting valves

MIDWAY ALONG A LINE of statues that overlooks I. M. Pei's glass pyramid at the Louvre, near the images of René Descartes and Voltaire, a rather forbidding figure looks down on the Napoleon Court. The man's right hand, as is traditional, is tucked into his coat. His left hand, howevcr, holds a curious contraption, something that looks a bit like a plumber's helper but is in fact one of history's most important leaps of mechanical imagination: the world's first steam-driven piston. The hand holding it belongs to its inventor, Denis Papin, whose ingenuity was critical to the creation of a steam-powered world, and whose life illustrates, as well as anyone's, the challenges of the inventive life.

The son of a government official in the city of Blois, Papin, a Huguenot (like many in the city, which had long been a haven for French Calvinists), was trained as a physician at the University of Angers and possibly even practiced as one for a few years, though his later comments suggest he much preferred physics. In 1671, he got the chance to act on that preference when he met the Dutch mathematician and physicist Christiaan Huygens, a founding member of the Académie des Sciences (inaugurated in 1666 as the French equivalent of the Royal Society), who was at Versailles repairing a balky windmill used to power the palace's fountains. The following year, Huygens, who had been impressed with Papin's mechanical insights, offered him a job as his secretary, and Papin gave up the healing arts for good, migrating to Paris to work at the Royal Library.

Huygens was another in a seemingly unending line of seventeenth-century scientists fascinated by vacuum and atmospheric pressure, and Papin's time with him was evidently both satisfying and productive. The two worked on a number of air pump experiments, and jointly published five papers in the *Philosophical Transactions* of the Académie Royale in 1675, though histories differ on whether they worked together on Huygens's gunpowder-driven piston, a promising but slightly hazardous technology.

During Papin's stay in Paris, life in France was becoming more than slightly hazardous for the nation's Huguenots, the beginning of a process that would end in the revocation of the Edict of Nantes and the return of official persecution, in 1685. By then, Papin had accurately read the writing on the wall, and, seeing no future for him in his birth nation, crossed to England in the fall of 1675. He was armed with a letter of introduction from Huygens to Robert Boyle, who was in need of a collaborator to replace Hooke, whose own researches were by then being financed by his employers at Gresham College and the Royal Society. The two evidently hit it off, and Papin joined Boyle as his secretary, though a better term would have been "experimental assistant."

While Papin was no Hooke (this is scarcely an insult: by 1675, Hooke had explained the twinkling of stars, described the earth's elliptical orbit, rebuilt the fire-destroyed Royal College of Physicians, disputed with Sir Isaac Newton over the discovery of the diffraction of light, and invented the anemometer, and he still had twenty productive years in front of him), he did excel at both experimental design and mechanical gadgetry. Most famously, in

1681 he invented a steam digester, or "machine for softening bones" as he described it, which was essentially a pressure cooker designed to clean bones rapidly for medical study.

The subsequent pattern of Papin's life would be familiar to any contemporary academic in search of a tenure-track position. In 1679, before the steam digester made him briefly famous, Papin was hired by his predecessor, Robert Hooke, as a secretary at the Royal Society at an annual salary of £20; he left there in 1681 for a new job as "director of experiments" at the Accademia Publicca di Scienze in Venice, yet another Royal Society imitator. After the Accademia failed, Papin returned to England, and Hooke, for three more years, this time as "Temporary Curator of Experiments" at the Royal Society, leaving *that* to become professor of mathematics at the University of Marburg.

Papin's contributions might have had an even larger impact had he enjoyed, like Boyle, the income from lands acquired by the Earl of Cork. And they are not small even so. In the 1686 issue of *Philosophical Transactions,* Papin describes (though evidently did not actually construct) an early air gun, probably a direct outgrowth of his gunpowder-piston experiments with his onetime mentor, Huygens. His digester featured a brilliantly innovative safety valve: When the pressure inside the chamber of Papin's invention grew high enough, it would overcome the weight of a hinged and weighted stopper and open a path to the outside, but when the pressure subsided, the stopper's weight would cause it to sink back to its normal position.

Most significantly for the evolution of the steam engine, in 1690 he published, in the *Acta Eruditorum* of Leipzig, a design of a true atmospheric engine: one that used the vacuum created by steam condensation to let atmospheric pressure drive a piston—the same one carried by his statue at the Louvre. Papin's great insight was recognizing that the weight of the atmosphere on the top of an open cylinder, which is apparent only when a vacuum is created at the cylinder's bottom, could also drive something mechanical *within* the cylinder. He wrote, "Since it is a property of water that a small quantity of it turned into vapour by heat has an elastic force like that of air, but upon cold supervening is again resolved in water, so that no trace of the said elastic force remains, I conclude that machines could be constructed wherein water, by the help of no very intense heat, and at little cost, could produce that perfect vacuum which could by no means be obtained by gunpowder."

By 1707, he was corresponding with Gottfried Wilhelm Leibniz, the German mathematician, engineer, and philosopher,* about the possibilities of an engine driven by steam pressure, all while trying to keep his head above water as a poorly paid councillor to Charles-August, Landgrave of Hesse-Kassel, a German principality located on the Prussian border. Keeping the landgrave interested proved a challenge all its own: Papin built him a centrifugal pump (evidently to water the landgrave's gardens) and a furnace air blower that became known as the "Hessian bellows." He even tried to design a hydraulic perpetual motion machine based on the belief that pressure from one large cylinder would provide a never-ending source of pressure on a smaller cylinder. By the time he built a demonstration submarine for his patron, however, the landgrave had already lost interest in it, and in Papin, who returned to England for the final time, spending his last years in unsuccessful attempts to promote a pension from the Royal Society and dying in poverty in 1712.

Papin was by all accounts a difficult man who lived a difficult life, and it is impossible to tell which was cause and which effect. He spent virtually all his adult years as a refugee, partly because of his religion—the late seventeenth century was no time to be a French Protestant—but even more because he was enormously rich in talents for which no market yet existed. He was an industrial scientist before there was an industry to employ him, which made him, in consequence, completely dependent on patronage.

* It is impossible to do justice to Leibniz with anything less than a full biography. He was simultaneously one of the greatest mathematicians and philosophers of the eighteenth century, with a list of achievements ranging from the calculus (the notation we use today is his, not Isaac Newton's) and binary logic to metaphysics, philology, and both basic and highly speculative physics.

His correspondence is evenly divided between generous sharing of his scientific discoveries and pleas for pensions, the latter wearing out his welcome in half a dozen countries. Papin's career, even more than Hooke's, illustrates the challenges faced by the most talented scientists if they lacked an independent source of income. The archetype—innovative talent supported either by patronage (governmental or aristocratic) or by inheritance—was as old as humanity and still quite sturdy.

Before he became an object lesson in the difficulties of making a living as a seventeenth-century inventor, however, Papin made one final connection on the route to engine 42B, and to *Rocket.* In 1705, Leibniz, then a courtier in the north German city of Hanover, received a sketch of a new machine for using steam to raise water, which he immediately sent to Papin in Hesse. The sketch had come from London.

THE TALLEST SKYSCRAPER IN the City of London, known variously as Tower 42 and NatWest Tower, occupies a site in Bishopsgate that was the former home of what was once London's only university: Gresham College, founded by a bequest from the will of Sir Thomas Gresham as a sort of scholarly Shangri-La, a college with neither students nor degrees. Instead, it houses scholars who offer lectures to any member of the public who cares to attend, and has been doing so ever since 1597. When Christopher Wren was tapped, in 1660, for the first lecture to what was to become the Royal Society, he was the Gresham Professor of Astronomy, and consequently that was where the lecture was given. The Royal Society called Gresham home for the next forty years, except for a brief period when fire and plague chased them out of London altogether.

Thus it was at Gresham College on June 14, 1699, that the Royal Society assembled for a demonstration of what was described as "a new Invention for Raiseing of Water and occasioning Motion to all Sorts of Mill Work by the Impellent Force of Fire which will be of great use and Advantage for Drayning Mines"—

in plain English, a steam engine. Its inventor was a military engineer named Thomas Savery.

The need for "drayning mines" was a relatively recent phenomenon, a direct consequence of the replacement of charcoal by pitcoal as the preferred fuel for space heating and for smelting metal. The preference was due less to the superiority of the mineral over wood, than to the fact that the raw material for charcoal was disappearing far faster than it could ever be produced. However, the deeper one digs for pitcoal, the greater the chance of finding water that needs "drayning," either by digging drainage tunnels, or *adits*—expensive, and only practical where the topography permits—or building pumps. The most powerful pumps in use in seventeenth-century England were operated by waterwheels, but nothing obliged rivers and streams to be convenient to mines; finding an alternative machine that could overcome water's tendency to seek the lowest level of any excavation meant that vacuum was no longer a purely philosophical concept.

Savery was not the first to realize that, just as turning water into steam created pressure, converting it back into water produced the opposite: a vacuum. By the middle of the seventeenth century, large numbers of people started to sense the enormous potential of a steam-created vacuum for pulling wealth out of the ground in the form not only of coal but also of copper, tin, and silver. Some of the attempts were made by Italians: in 1606, a Neapolitan engineer named Giambattista della Porta designed a machine to pump water out of a closed container using steam; some by Frenchmen: in 1609 or 1610, Salomon de Caus, an ambitious gardener who specialized in designing fountains, traveled from Dieppe to England, where he built a number of steam-driven toys at one of the residences of the Prince of Wales. And some were Englishmen, like the now forgotten David Ramsay, who supposedly invented, in 1631, a device "to Raise Water from Lowe Pitts by Fire," or the Marquess of Worcester, whose "water-forcing engine" dates from 1663.

The inability of della Porta, de Caus, and others to produce a working steam pump was, in some sense, as valuable as success might have been, since they failed publicly enough that others were able to learn from their failures. Thomas Savery was one of them, and it is worth noting that his own experiments were financed not by a wealthy aristocrat, but by a national government.

This is a poorly understood aspect of the Industrial Revolution. It doesn't fit very well with either a heroic entrepreneurial history in which visionary innovators, usually working alone, develop the ideas, machines, and institutions of progress, or a deterministic one, in which technological progress is a function of predictable natural laws. The messy truth turns out to be that the innovative culture that blossomed in eighteenth-century Britain depended both on individuals looking out for their own interests, and on recognizing a national interest in innovation. When Savery started investigating the "impellent force of fire," he was almost certainly working on his own behalf. But he did so at an English government facility: the Royal Office of Ordnance, which supported a large number of workshops and factories around London, and whose sole purpose was improving the technology of war. And so they did; though the cannon of the era were still mostly manufactured by private contractors located in the Weald, an ironworking center forty miles south of London, the Office of Ordnance tested them, and, more significantly for an engineer like Savery, "was responsible for the design and fabrication of various military engines . . . cranes, devices for mechanically hurling projectiles, gun carriages . . . and pontoon bridges for spanning streams." Sometime around 1639, the original Lambeth works of the Office of Ordnance had been expanded to include part of an ancient estate known variously as Fauxhall or Vauxhall, and made "a place of resort for artists, mechanics, &c [where] experiments and trials of profitable inventions should be carried on"—a sort of seventeenth-century equivalent of the U.S. Department of Defense Advanced Research Projects Agency, or DARPA, whose self-described mission is "to prevent technological surprise to the U.S. [and] to create technological surprise to our enemies."*

As with DARPA—which is where, among other things, the predecessor of the Internet was invented—engineers at Vauxhall produced technological surprises for the civilian world as well as the military. British monarchs, after all, had interests in mining as well as conquest, so it is no coincidence that one of Savery's predecessors at Vauxhall, Samuel Moreland (or Morland), an engineer in the employ of Charles II, made some sort of fire-driven water pump, "a new invention for raising any quantity of water to any height by the help of fire alone," in 1675. Moreland left behind not only his notes about the pump (since vanished) but something far more useful: a calculation of the volume of steam—about two thousand times that of water.

Moreland's calculation was not only the most precise estimate for more than a century; it was also critical for building any sort of working steam engine. Knowing that steam, once condensed back into water in a sealed container, leaves behind a vacuum that takes up two thousand times the cubic area of the condensed liquid is very nearly as important as knowing the temperature at which water boils—perhaps more so, since the first working steam engines were built decades before the first accurate thermometers.

Thus, when Savery made the first demonstration of his pumping machine "at a potter's house in Lambeth," two years before he did so with an identical machine at Gresham, he owed a large debt to his employers at the Royal Office of Ordnance. That machine consisted of a tall cylinder filled with water and connected to a boiler, in which Savery produced steam, which he then intro-

* Even more famous, and earlier, was the arsenal of the Republic of Venice, founded in 1104, which covered sixty acres and employed more than a thousand artisans, and which inspired even Galileo, who opened the *Dialogue Concerning Two New Sciences* by complimenting the Venetians whose "famous arsenal suggests to the studious mind a large field for investigation, especially that part of the work which involves mechanics, for in this department all types of instruments and machines are constantly being constructed by many artisans."

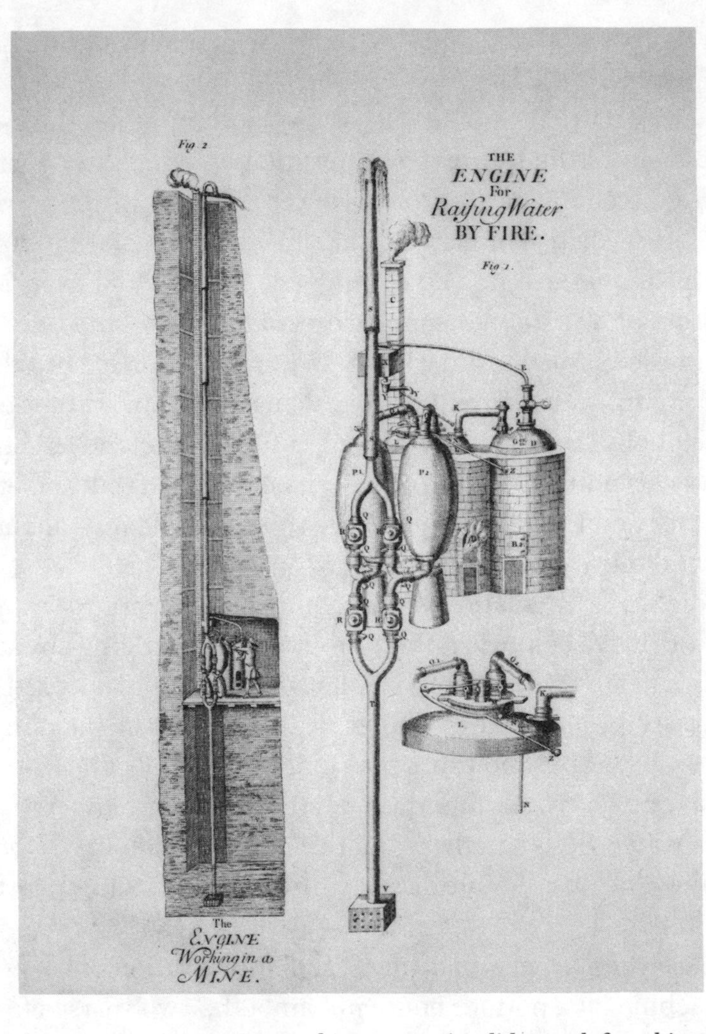

Fig. 1: Thomas Savery's pumping machine, as seen in a lithograph from his 1702 book *The Miner's Friend*. The image on the right shows the components: When the canister sprayed cold water on the steam filled cylinder, the resulting vacuum pulled the water up. On the left is the machine at work, two-thirds of the way down a mine shaft, since the vacuum could pull the water only a bit more than twenty feet. *Science Museum / Science & Society Picture Library*

duced into the cylinder at a pressure estimated to be about 120 pounds per square inch. The pressure pushed the water out one end of the cylinder, leaving steam behind; when the steam-filled cylinders were sprayed with cold water, the resulting vacuum pulled water from a chamber below, creating a pumping action.

Savery's machine was a long way from perfect. The use of water as its "piston" meant that the engine couldn't pump anything else. Unlike Papin's digester, it lacked any sort of safety valve. Using it even in its (slightly) improved version would have required an operator to open the steam cock and the cold water valve at least four times a minute; to refill the boiler at least once a minute; and to stoke the fire under it as needed. The cylinder was soldered together, and the solder had a melting temperature dangerously close to the temperature of the steam at high pressure, which could exceed 350°F/175°C. Worst of all, while the high-pressure steam could, in theory, push the water several hundred feet upward, the machine depended on suction for the first leg of the journey. Any working Savery pump needed to be built not at the top of a mine shaft, but no more than twenty-five feet from the bottom.

The reason for this limit, which had been well known to Galileo, Torricelli, and Papin, was atmosphere. At sea level, the maximum height that water can be lifted inside a tube under perfect conditions is about thirty-four feet, or just a little more than ten meters, calculated by dividing atmospheric pressure at sea level (14.7 pounds per square inch) by the weight of a cubic inch of water, or 0.0361 lbs. At this point, the water inside exerts a pressure equal to the weight of the atmosphere pushing down on the water's surface. Since this represents a theoretical limit, requiring a perfect vacuum, the practical limit is even lower, usually assumed to be around twenty-five feet, as anyone who has tried to use a suction pump to draw water to the top of a three-story building has learned. This was a fairly serious problem for draining mines that were already more than one hundred feet underground.

Nonetheless, Savery's machine was a revelation. And not merely to the "gentlemen, free and unconfin'd" of the Royal Society, who reported, with characteristic understatement, "Mr. Savery . . . entertained the Royal Society with shewing a small model of his engine for raising water by the help of fire, which he set to work before them, the experiment succeeded according to expectation, and to their satisfaction," but to King William III in a private showing at Hampton Court. More modestly, but far more importantly, it also inspired a Devonshire ironmonger and blacksmith named Thomas Newcomen.

FOR CENTURIES, THE LANDED gentry who held the lands around Dudley Castle in the West Midlands of England prospered in direct proportion to the value of the minerals extracted from those lands. Indeed, that prosperity often took precedence over maintaining the land, and by the mid-1660s, the current Baron Dudley had heavily mortgaged the lands—so heavily, in fact, that he was forced to marry his daughter to someone wealthy enough to pull the family out of debt. The priority of succeeding barons was, as a result, the revitalization of the Dudley real estate, the most valuable pieces of which were the Conygree coal mines, lying one mile east of Dudley Castle.

The Conygree mines, like all excavations, were only workable when dry, or at least free of standing water. This, of course, is why Savery called his pump the "Miner's Friend" in an eponymous 1702 book. The book, and the invention, demonstrated how Torricelli's (and von Guericke's) vacuum could be economically created using the two-thousandfold difference in volume between water in its liquid and gaseous state, and showed how such a vacuum could pull water out of any mine.

So long as the pump could be built no more than twenty-five feet from the mine's floor.

After two hundred years of excavating, however, the mines at Conygree were more than six times deeper than the working distance of a vacuum pump, which meant a Savery-style engine would need to be built (and operated) more than one hundred and twenty-five feet below ground level. What they needed was an entirely new machine. Even more, they needed an act of genius, and this time the word, so frequently devalued by overuse, is appropriate. One historian of science calls the machine that made history near Dudley Castle in 1712 "one of the great original synthetic inventions of all time."

The synthesis in question tied together two intellectual threads whose history dates from Heron's first-century Alexandria. The first was man-made vacuum: the concept that was studied by Torricelli and pursued by Giambattista della Porta and Salomon de Caus, and that reached its culmination in Savery's "new Invention for Raiseing of Water." The other was the realization that a functional piston could be driven by atmospheric pressure, which was investigated by Huygens and described, though not built, by Papin in 1690. The 1712 engine of Thomas Newcomen, probably the first working engine built by this enigmatic man, was certainly the first to connect the two threads; and if any single invention can be said to have inaugurated the steam revolution, this was it.

Much more is known about the machine than its inventor. He was born in 1664 in Dartmouth, to a family that may have been in the shipbuilding trade. In his teens he was in all likelihood apprenticed to an ironmonger—part smith, part hardware salesman—since he was practicing the trade as a journeyman by the age of twenty-one, but his name doesn't appear in records of indentured freemen, likely because his family was Baptist in a land that recognized only the Church of England.

Newcomen's religion had consequences greater than absence from a local census. Dissenters, including Baptists, Presbyterians, and others, were, as a class, excluded from universities after 1660, and either apprenticed, or learned their science from dissenting academies.

Bad luck for the universities, good luck for the nation. Only decades after a tidal wave of scientific knowledge started washing over Britain—the first English translation of Galileo's *Dialogue Concerning Two New Sciences* was published in the 1660s, nearly seventy years before an English edition of the *Principia* of Isaac Newton (Latin edition, 1687)—some of the nation's most ambitious and practical young were excluded from Oxford and Cambridge. At the same time that he chartered the world's first scientific society, Charles II had created an entire generation of dissenting intellectuals uncontrolled by his kingdom's ever more technophobic universities. Some attended so-called dissenting academies, which mimicked an Oxbridge classical education with notably less arrogance about the teaching of science and modern languages. Many more learned their science in the most practical way: as apprentices to artisans who were more likely to be literate than ever before in history.

Newcomen may have been unable to translate Horace, but that did not mean he was, in any important sense, uneducated. He could perform calculations rapidly, knew a fair bit of geometry, could calculate the strength and velocity of moving parts, could draw clearly, and—obviously—could read all that was available on subjects that interested him.

With books to read, and tools to practice his trade, Newcomen might still have lacked sufficient resources to travel all the way to Conygree but for one more unanticipated consequence of the Restoration: a package of laws that prohibited pastors who refused to conform to their dictates—using the Book of Common Prayer, for example—from teaching or preaching anywhere within five miles of their former "livings." As a result, Dartmouth's Baptists hired as their pastor the Reverend John Flavel, a well-known Presbyterian who not only led secret community services (another law forbade religious gatherings of more than five people) but organized secret community banks as a method of pooling their resources. One of them funded Newcomen's first experiments.

This is worth underlining. The significance of the Dartmouth "bank" in the history of steam power is real, but modest. However, it is also a reminder of what we might call the British advantage in the development of the Industrial Revolution. Compare, for example, the experience of being a Baptist in Restoration England with that of being any sort of Protestant in seventeenth-century France. Newcomen may have been invisible to his local census, but at least he was not, like Papin, exiled from his country.

Thus, between 1700 and 1705, while Papin was wandering across central Europe trying to secure a pension, Newcomen, and his partner, John Calley (sometimes Cawley), a glazier (sometimes a plumber), set up a workshop in Newcomen's basement, financed by Flavel's bank, and started experimenting. During the next decade, more or less, most of their time was spent on trying to improve one or the other of the two seemingly independent threads of steam engine development: Savery's vacuum, and Papin's piston.

Frustratingly, we know little of just how and when the knowledge of the two came to Newcomen. Relatively detailed descriptions of Savery's "Miner's Friend" were, of course, available to anyone who could read once it appeared in the Royal Society's *Philosophical Transactions* in 1699, and certainly after the one-time military engineer published his book cum sales brochure in 1702. Intriguingly, Savery was then living in Modbury, only fifteen miles from Newcomen's Dartmouth home, and we know that he was regularly hiring artisans to build models and parts for his engines. Given Newcomen's reputation as an ironmonger and wheelwright, it isn't a huge leap to imagine some contact between the two.

Details about Papin's piston-driven engine weren't quite as public, but they were scarcely secret. Newcomen maintained active correspondences with a number of contemporaries, none more important than that with Robert Hooke, one of the most wide-ranging intelligences of the entire century. Hooke corresponded not only with Newcomen but with Papin as well, and he very likely kept the former apprised of the latter's progress. At some point before his death in 1703, Hooke even talked Newcomen out of Papin's idea of driving the pump's pistons by air pressure, urging him instead to pursue the idea of creating a vacuum under the piston, writing "could he [Papin] make a speedy vacuum under your piston, your work is done. . . ."

Well, not quite done. Newcomen's real conceptual breakthrough came when he finally combined the strongest features of the two different approaches—or, at least, discarded their weaknesses. His brilliant synthesis lay in forgoing Savery's dependence on vacuum to raise water, and Papin's use of a piston operated by expanding steam. And in adding one critical element: the beam.

The most conspicuous mechanical element in Newcomen's 1712 engine—for that matter, the most conspicuous element in virtually every steam engine for the next century and a half, including the Crofton pump station's engine 42B—was its horizontal working beam. It looks a bit like an unbalanced seesaw, with the underside of one end attached to a piston and the other to a pump rod holding a bucket, which made the pump end much heavier. When at rest, therefore, the beam angled down toward the bucket at the bottom of the mine shaft, which forced the piston, inside a cylinder filled only with air, up to its highest point. Since the bucket on the end of the pump rod could be hundreds of feet below the guts of the engine, the critical problem with Savery's engine—the need to place it near the bottom of the mine shaft—was solved. In fact, the only limitation on the depth at which it could work was the weight of the cable holding the bucket, which was relatively insignificant.

Newcomen's first brilliant innovation—to lift water by seesawing a horizontal beam—was entirely dependent on his feel for the machine's geometry; the beam, however, didn't do anything about the need to renew the cycle of vacuum in a "speedy" manner, as advised by Hooke. This was critical for a working machine, which had to do more than impress German princelings or even the Royal Society; it had to return to room temperature after being heated well past the boiling temperature of water, and it had to so a dozen times a minute.

Savery's method for producing condensation—spraying cold water on the outside of the cylinder—was simply too slow. Calley and Newcomen had designed a lead envelope to surround the cylinder, into which cold water could be poured, which improved the speed for heating and cooling the cylinder, but not a lot. Enter luck: The cylinder was essentially a flat piece of tin wrapped into a cylinder shape, its ends held together with a strip of solder. At one point, the solder was imperfectly applied, and the heat of steam in the cylinder melted it, opening a hole. When Newcomen poured cold water into the lead envelope wrapped around the cylinder, a stream of it found the hole, rushed through it, and condensed the steam immediately, with powerful results. For purposes of the experiment, Newcomen had attached a weight to the end of the beam to represent the weight of water; when the steam condensed, it pulled the beam down so violently that it broke the chain, the bottom of the cylinder, and even the lid of the boiler underneath.

Newcomen and Calley had, in broad strokes, the design for a working engine. They had enjoyed some luck, though it was anything but dumb luck. This didn't seem to convince the self-named

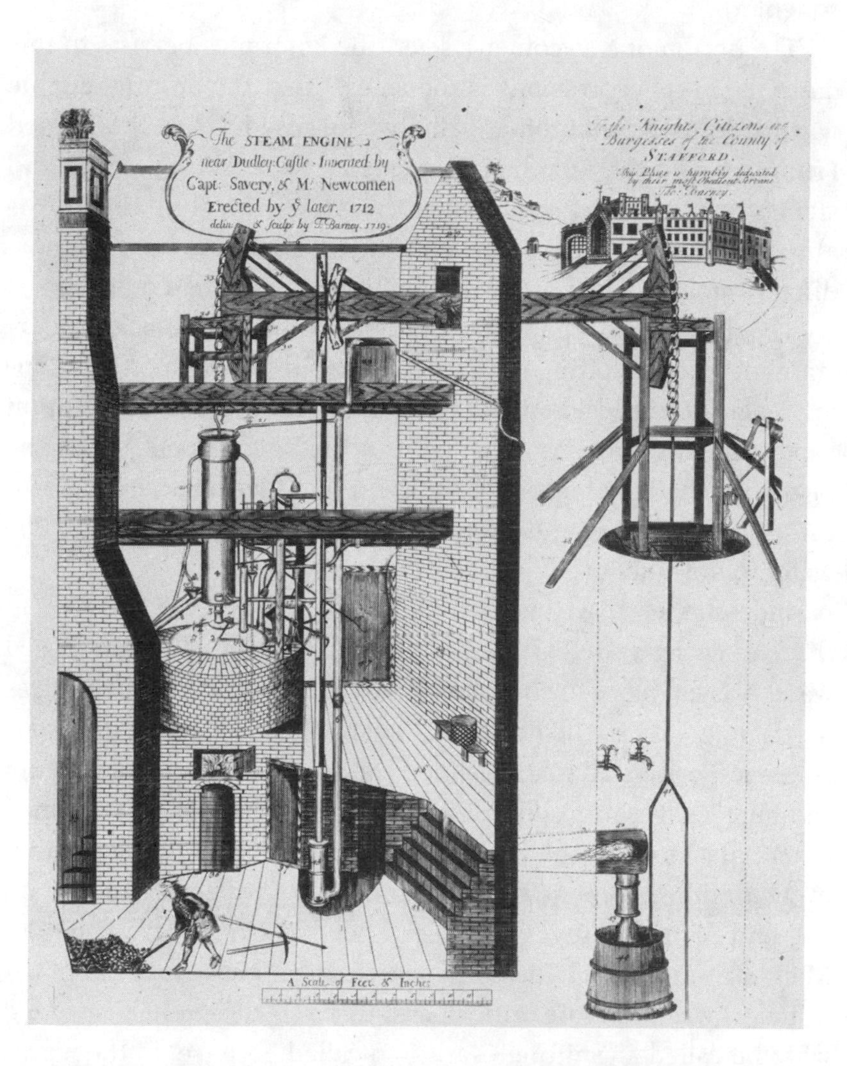

Fig. 2: The engine that Thomas Newcomen and John Calley erected at Dudley Castle in 1712, as seen in a 1719 engraving, used its vacuum to drive not water, but a piston attached to a beam. *Science Museum / Science & Society Picture Library*

experimental philosopher J. T. Desaguliers, a Huguenot refugee like Papin, who became one of Isaac Newton's assistants and (later) a priest in the Church of England. Desaguliers wrote, just before his death in 1744, that the two men had made their engine work, but "not being either philosophers to understand the reason, or mathematicians enough to calculate the powers and to proportion the parts, very luckily by accident found what they sought for."

The notion of Newcomen's scientific ignorance persists to this day. One of its expressions is the legend that the original engine was made to cycle automatically by the insight of a boy named Humphrey Potter, who built a mazelike network of catches and strings from the plug rod to open the valves and close them. It is almost as if a Dartmouth ironmonger simply *had* to have an inordinate amount of luck to succeed where so many had failed.

The discovery of the power of injected water was luck; understanding and exploiting it was anything but. Newcomen and Calley replaced the accidental hole in the cylinder with an injection valve, and, ingeniously, attached it to the piston itself. When the piston reached the bottom of the cylinder, it automatically closed the injection valve and opened another valve, permitting the water to flow out.

Indeed, the valves, one between the boiler and the piston, and the self-acting valve, with a flap that closed once the condensed water was let out of the bottom of the cylinder, demanded quite as much ingenuity as the horizontal beam itself. All of the discoveries since Torricelli had underlined the potential power of vacuum combined with atmospheric pressure, but the power remained potential so long as the vacuum was unstable; losing the vacuum in the middle of a cycle was functionally equivalent to getting off one end of the seesaw while the other kid is still up in the air. Maintaining it, which meant in practice keeping any air out of the cylinder, was therefore critical, and to do so Newcomen invented what he called a "Snifting Clark" (so called because, in the words of a contemporary observer, "the Air makes a Noise every time it blows thro' it like a Man snifting with a Cold"), a valve carefully designed—not too heavy, not too light—to blow the air out of the chamber without letting any steam escape. Another vacuum preserver, probably the simplest, was the layer of water Newcomen added at the top of the piston, which served to seal the chamber from any air, which would compromise the vacuum.

Newcomen spent ten years experimenting with solutions to the problem of maintaining a regular and stable motion in his engine. None of his solutions was more innovative than his so-called plug rod. Since the machine depended on regular injections of water to condense the steam, it required an equally regular water supply. In Newcomen's machine, this water was held in an overhead tank; gravity could be relied upon to move water from the tank into the cylinder, but to feed the tank itself, another pump was necessary. Newcomen suspended the plug rod from the horizontal beam itself; this rod, in turn, operated the cylinder valves, thus connecting the flow of water from one chamber into another. As water was pumped into the overhead tank, it also lifted the plug rod and thereby opened the valves of the cylinder, giving the beam a continuous (though jerky) motion. The ingenious F-shaped lever that opened the catch and operated the injection valve may have been a primitive design, but when the University of Manchester Institute of Science and Technology built a scale model of the original 1712 engine in 1968, "the valve still functioned perfectly, and was another amazing case of Newcomen arriving at the correct answer."

Even more elegantly, the onetime ironmonger designed a Y-shaped lever to control the steam entering the engine itself. The lever stayed balanced on the trunk of the "Y" until the piston reached the bottom of its stroke, where an attached peg pushed one of the arms, overbalancing and opening the steam valve, simultaneously destroying the vacuum and pushing the air out through the snifting valve. As the piston rose, the valve stayed open until the top of the stroke, when another peg pushed the other arm, shutting the valve during the complete working stroke.

Newcomen's valves aren't just expressions of how well he had trained his mind during his years of experimentation. They also tell of the years he spent "educating" his hands at the blacksmith's anvil, the mechanic's lathe, and the carpenter's bench.

This insight is hugely important for understanding not only invention generally, but the era of sustainable invention that Thomas Newcomen inaugurated. Consider, for example, a single element of the Newcomen design: the Y-valve. It was utterly essential to the stable functioning of the engine, but to do its job, it needed to be precise both as to shape *and* to weight; it would only work if it rocked back and forth on its base as the piston rose, which meant that it needed to be balanced with exactly the same mass on each "arm" of the Y.

Now imagine producing such a fitting when the only tools available were a hammer, a chisel, and a file, and perhaps a set of calipers for measurement; no lathes (at least, no lathes that could work metal), no drills, and certainly no powered tools, all of which were decades in the future. The *only* way to "machine" such a valve was by hand, and the hands in question had to be as sensitive, and as precise, as those of a violinist. Newcomen could perhaps imagine the shape of his valves by eye, but he needed to feel their weight, and their texture, with his hands.

For centuries, certainly ever since Immanuel Kant called the hand "the window on the mind," philosophers have been pondering the very complex way in which the human hand is related to the human mind. Modern neuroscience and evolutionary biology have confirmed the existence of what the Scottish physician and theologian Charles Bell called "the intelligent hand." Stephen Pinker of Harvard even argues that early humans' intelligence increased "partly because they were equipped with levers of influence on the world, namely the grippers found at the end of their two arms." We now know that the literally incredible amount of sensitivity and articulation of the human hand, which has increased at roughly the same pace as has the complexity of the human brain, is not merely a *product* of the pressures of natural selection, but an *initiator* of it: The hand has led the brain to evolve just as much as the brain has led the hand. The hands of a pianist, or a painter, or a sushi chef, or even, as with Thomas Newcomen, hands that could use a hammer to shape soft iron, are truly, in any functional sense, "intelligent."

This sort of *tactile* intelligence was not emphasized in A. P. Usher's theory of invention, the components of which he filtered through the early twentieth-century school of psychology known as Gestalt theory,* which was preeminently a theory of *visual* behavior. The most important precepts of Gestalt theory (to Usher, anyway, who was utterly taken with their explanatory power) are that the patterns we perceive visually appear all at once, rather than by examining components one at a time, and that a principle of parsimony organizes visual perceptions into their simplest form. Or forms; one of the most famous Gestalt images is the one that can look like either a goblet or two facing profiles. Usher's enthusiasm for Gestalt psychology explains why, despite his unshakable belief in the inventive talents of ordinary individuals, he devotes an entire chapter of his magnum opus to perhaps the most extraordinary individual in the history of invention: Leonardo da Vinci.

Certainly, Leonardo would deserve a large place in any book on the history of mechanical invention, not only because of his fanciful helicopters and submarines, but for his very real screw cutting engine, needle making machine, centrifugal pumps, and hundreds more. And Usher found Leonardo an extraordinarily useful symbol in marking the transition in mechanics from pure intuition to the application of science and mathematics.

But the real fascination for Usher was Leonardo's straddling of two worlds of creativity, the artistic and the inventive. No one, before or since, more clearly demonstrated the importance to invention of what we might call "spatial intelligence"; Leonardo was not an abstract thinker of any great achievement, nor were his mathematical skills, which he taught himself late in life, remarkable.

* Gestalt, for those preparing for a midterm on the history of cognitive science, was an attempt to explain, or at least describe, the way in which the mind integrates perception with cognition, developed by Germans Max Wertheimer, Wolfgang Kohler, and Kurt Koffka, who were in turn strongly influenced by the work of the Austrian physicist Ernst Mach, who will crop up again in chapter 8.

His perceptual skills, on the other hand, developed primarily for his painting, were extraordinary; but they were so extraordinary that Usher could write, "It is only with Leonardo that the process of invention is lifted decisively into the field of the imagination. . . ."

Seen in this light, Usher's attention to Leonardo makes perfect sense. What the great artist-inventor "saw," in Gestalt terms, was determined as much by what was inside his head as by what was in front of his eyes. Leonardo's gifts, his education, and his history constrained his perceptions, but also gave them direction. Any inventor's moments of insight, certainly including Newcomen's and supremely Leonardo's, are primarily visual; as a modern scholar puts it, "Pyramids, cathedrals, and rockets exist not because of geometry, theory of structures, or thermodynamics, but because they were first a picture—literally a vision—in the minds of those who built them . . . technology has a significant intellectual component that is both nonscientific and nonliterary."

There is, however, something missing from Usher's pure Gestalt explanation of the process of invention. For while it seems reasonable to suppose that most insights are visual, the equally creative process of critical revision is almost overwhelmingly tactile. Leonardo's hands—holding a brush or a pen, or building a model—were as important as his eyes.

This had obviously been true for a thousand generations of craftsmen and artists, including Praxiteles, Mozart, and, once again, Leonardo. But with the beginning of the eighteenth century, the implications changed, and the reason was an unprecedented enthusiasm for scale models.

The eighteenth century's need for mechanical models that could be enlarged by orders of magnitude while still performing as they did in their miniaturized form has no precedent in history. Before then, except for the work of a few outliers like Leonardo, mechanical objects that were built by "intelligent hands" were usually as large as they were ever going to get; only with the advent of scale modeling, frequently performed in stages during which a device could grow from the size of a suitcase to that of a house in several steps, were those hands employed in improving mechanisms bigger than toys, or, as was the case with Hooke and Boyle, scientific apparatuses. Thereafter, the critical revisions that Usher described were going to be performed by, and improved by, the hands of trained artisans. The intelligent hands of Thomas Newcomen made him eighteenth-century England's first important craftsman-inventor. He would not be the last.

This was because Usherian critical revision is a social process, in which the insight of one inventor is revised and reinforced by others. The inventor, in Usher's words, "lives in the company of a great company of men, both dead and living." By the beginning of the eighteenth century, a literate artisan class, trained in practical mathematics and engineering, was exhibiting a never-before-seen passion for revising and reinforcing one another's inventions. The size of that "great company of men" and therefore the potential for cross-fertilization—for critical revision—had exploded.

SO IT WAS THAT while only Thomas Newcomen and John Calley arrived at the Conygree mine a mile east of Dudley Castle that day in 1712, a "great company of men" from different times and places accompanied them, observed the construction of the sturdy machine on site, and (metaphorically) applauded when the boiler was fired up, the steam was injected into the cylinder, and the beam rocked on its pivot for the first time. Every stroke of Newcomen and Calley's engine lifted ten gallons of water out of a 50-meter-deep mine, and did so at a rate of twelve strokes per minute. "During the up-stroke, the water was drawn up into the pump cylinder as the bucket [at the end of the connecting rod] was being raised. Valves in the bottom of the bucket opened on the down-stroke to let the water pass through to the upper side . . . and was then lifted on the next up-stroke."

The inventive power is inherent in the idea that a column of air has weight in the same way as does a column of bricks, and that if some "bricks" of air are removed from the bottom of the column, the top will move downward. Newcomen's cylinder removed the bricks by condensing steam into water, under the watchful, if figurative, eyes of Galileo, Torricelli, Boyle and Hooke, Denis Papin, Heron of Alexandria, Otto von Guericke, and a thousand others. What was on display was, even by their elevated standards, genius.

Yet there is an even more important way in which the engine on display at Conygree in 1712 was a work of genius. In its original Latinate meaning, the word is defined as the guardian spirit of a particular household, or tribe: a *gens;* and the genius responsible for the machine at Conygree was, indeed, from a very particular tribe. In Newcomen's case, he was the first (or very nearly) and clearly the most important member of a tribe of a very particular, and historically original, type: the English artisan-engineer-entrepreneur. His great invention was not merely an ingenious toy but a profitable one, whose great virtue was not only its productivity but its simplicity and ruggedness. Like another century's AK-47 assault rifle, it survived the ministrations of even the most technically inept users. It was forgiving, requiring no real precision in alignment, and could be built (except for the cylinder) by local craftsmen using local materials. It truly deserves its description, given by Abbott Payson Usher himself, as "the greatest single act of synthesis in the history of the steam engine, and must be regarded as one of the primary or strategic inventions" of all time.

Unfortunately for Newcomen, one of the "company of great men" present at Conygree in spirit had a legal interest in the proceedings. Thomas Savery would live only three years after Newcomen's demonstration, but his patent survived. What this meant for Thomas Newcomen was that in order to exploit his invention—and he clearly wanted to: in a 1722 lawsuit, he described himself from the beginning as "designing to turn his engines or part of them into cash"—he was compelled to make his peace with Savery, or rather, Savery's heirs. There was little question that the 1712 engine represented an enormous advance over the 1698 "Miner's Friend"; the Savery engine had found only a few customers, and of those he had installed, most, including the engine used by the York Buildings Company to pump water to its London customers and the one used to drain the Broad Waters pool in Staffordshire, showed a disconcerting tendency to blow up. But Savery still owned an exclusive on the concept, and (because of a special Act of Parliament) would continue to do so until 1733. This was eighteen years after Savery's death, but his exclusive rights survived. And so, therefore, did a method for earning money from those rights; at Savery's death in 1715, his partners created a company named "The Proprietors of the invention for raising water by fire." According to an adviser to one of those Proprietors, the lawyer Sir Thomas Pengelley, Savery "divided the profit to arise by his invention into 60 shares . . . after his death, his executors sold the rest. One Mr. Newcomen having made considerable improvements to the said invention, the Proprietors in 1716 came to an agreement amongst themselves and with the said Newcomen and by indenture in pursuance of articles made between Savery and Newcomen, they agreed to add 20 shares to the 60 which were to be had by Newcomen in full of his improvement and of the said agreement."

THOUGH NEWCOMEN'S TAKE FROM the sale of the engines was "only" one-quarter—twenty shares out of eighty—they were enough to guarantee his prosperity, since the new company charged up to £300 a year just to license the machine; one coal miner paid the Proprietors £200 plus half his yearly profits—and that was in addition to paying for building the engine itself. The year the company was founded, an article appeared in the *London Gazette* with the following teaser:

> Whereas the invention for raising water by the impellent force of fire, authorized by Parliament, is lately brought to the greatest perfection, and all sorts of mines, &c., may be thereby drained and water raised to any height with more ease and less charge than by the other methods hitherto used . . . these are therefore to give notice that if any person shall be desirous to treat with the proprietors for such en-

> gines [i.e. Newcomen and Calley; all the examples given in the article are of their installed engines] attendance will be given for that purpose every Wednesday at the Sword Blade Coffee House in Birchin Lane, London.

Within three years, more than a hundred Newcomen-style engines were pumping away in various parts of England, all of them helping the onetime ironmonger and Baptist lay preacher to "turn his engines into cash." He owed his success in doing so to predecessors making discoveries as far away as Tuscany and as close as Oxford; but the most important occurred only about three miles west of Birchin Lane, on the banks of the Thames, in sight of Westminster Abbey.

CHAPTER THREE

THE FIRST AND TRUE INVENTOR

concerning a trial over the ownership of a deck of playing cards; a utopian fantasy island in the South Seas; one Statute and two Treatises; and the manner in which ideas were transformed from something one discovers *to something one* owns

IN THE YEAR 1602, what would be the last full year of the reign of the first Elizabeth, her kingdom's capital was one of the world's largest and most vibrant cities, but it would be largely unfamiliar to the modern Londoner. A century before Newcomen began crafting the world's first steam engine, St. Paul's Cathedral was still awaiting the fire that would consume its predecessor, and the consequent opportunity to rebuild it. No Buckingham Palace. Hampton Court might be half-familiar, like a poorly remembered dream; most of the modern incarnation was rebuilt after the fire by Christopher Wren for its new owner-occupants, William and Mary. Moreover, for every building in modern London not yet built in the year 1602, there is another from 1602 that no longer stands.

Of the Houses of Parliament, only Westminster Hall remains. Built by William Rufus, son of the Conqueror, and rebuilt by Richard II, who added the great oak hammer-beam roof sometime between 1395 and 1399, it still impresses today, partly because of its scale. The hall was, and is, a huge space, two hundred and fifty feet long by seventy feet wide, with nothing to divide its many functions one from the other—a nontrivial characteristic for a place that England's monarchs were as likely to use to play tennis as to announce vital decisions about state policy.

Even so, the really impressive aspect of Westminster Hall was that it was, until the 1673 construction of the Old Bailey, the most important courtroom in the realm. This was where William Wallace was tried and found guilty of treason in 1305. In January 1606, it was the site for the trial of Guy Fawkes, Robert Winter, and the other conspirators in the Gunpowder Plot; and, most famously of all, it was the place where, in 1649, Charles I was charged and found guilty of treason and other high crimes. Less well known, but just as historically significant—more significant, from the standpoint of *Rocket*—is the case sometimes recalled as the Case of Monopolies, the lawsuit that marks the ideological transformation that would, decades hence, create the Industrial Revolution. In one of history's odder twists, some of the track on which *Rocket* would one day ride was laid by a lawyer.

Not just any lawyer, to be sure. His name was Edward Coke (pronounced "cook"), and he was, as the sixteenth century turned into the seventeenth, the most prominent, successful, and honored lawyer in England. Before he turned forty, Coke had served as a Member of Parliament, as the Speaker of the House of Commons, and as England's Attorney General; by the early 1600s, his reputation, already enormous, reached rarefied heights—or at least the height of the Tower of London, the site of the execution of the Earl of Essex, prosecuted by Coke for treason against Elizabeth, and the imprisonment of Walter Ralegh for treason against her successor, James I.

Coke had also, by the time of the accession of James, acquired more than a reputation. Though already in possession of an income estimated at £12,000 a year, as large as any in the kingdom, he hatched a plan to solicit "every man of estate" to sue the incoming monarch for a pardon, for the fee of £5 a head. Apparently enough prominent Englishmen were worried about the accession of the first non-Tudor king in more than a century that staggering numbers of them bought into what looks suspiciously like a protection racket; at least one estimate puts Coke's take at £100,000.

Coke may not have come by his wealth honestly, but he earned his reputation as the most influential English jurist of all time. His 1628 "Petition of Right," which enumerated the limits on the power of the king, is not only one of the so-called "Fundamental Laws" of England, but was a precipitating cause of the Civil War. However, neither Coke's great wealth, his brilliant forensic talent, nor his centuries-long influence on Anglo-American political institutions explain his presence in the middle of a history of the steam revolution.

On the other hand, *Darcy v. Allein,* the proper name for the litigation known as the Case of Monopolies, does.

In 1602, when the Case of Monopolies was first presented, Coke was fifty years old and had been a practicing barrister* for twenty-four years, representing both the sovereign and a dizzying number of other clients including "country gentlemen, acquisitive parsons, Roman Catholic exiles, puritan dissidents, cockney pub licans and City haberdashers, duelists, forgers, and burglars." Many of the most familiar portraits of Coke were painted in his old age and show a rather spare, elegant man with sunken cheeks and a carefully groomed and pointed beard, but the fifty-year-old version was still a big man, made bigger by his robes, with a large head, wide shoulders, famously piercing eyes, and a penchant for intimidating witnesses and opposing counsel alike with his size, his aggressive manner, and his idiosyncratic language.

The law in England is, after all, an idiosyncratic profession, with

* In different parts of the world, the two functions of legal professionals—advising on the law and advocating before judges—are either split or fused. In the English tradition, the first function was traditionally performed by professionals known as solicitors, the latter by barristers, so named because of the literal bar that separated students from practitioners in the Inns of Court. In the United States, and increasingly in the United Kingdom (even to the point of permitting solicitors to wear powdered wigs), the functions are performed by the same lawyers.

its own rituals, language, and history. In order to practice it, for example, a barrister was required to affiliate with one of the four so-called "Inns of Court"—Lincoln's Inn, Gray's Inn, the Middle Temple, and the Inner Temple—which offered room and board to students and, effectively, a license to practice to barristers, who were otherwise barred from pleading cases in court. Coke's home base at the time of the Case of Monopolies was the Inner Temple, part of a compound originally built in the twelfth century as a church and residence for a monastic military order founded by two First Crusade refugees, and known as the Poor Soldiers of Christ and the Temple of Solomon, more popularly the Knights Templar.

In Coke's time, even the judicial calendar was idiosyncratic; *Darcy v. Allein* was to be heard during the Michaelmas term, one of Westminster Hall's four sessions, set to the rhythm of the Christian festival year: Michaelmas (early October to late November), Hilary (from late January to late February), Easter, and Trinity (the two weeks after Trinity Sunday). To this day Oxford University's academic year uses the same terminology, for no explainable reason. We can therefore infer that as *Darcy v. Allein* came to trial, Westminster Hall was cold* but still warmed by the thousands of litigants, visitors, and lawyers that crowded the space all year long.

Crowded, and noisy. Westminster Hall might have been nearly the size of a football field, but it had to accommodate dozens of cases, with hundreds of lawyers arguing with one another, all at the same time. No walls separated the Court of Common Pleas, for example, which heard most civil suits (then and now, the bulk of the work of any judicial system), from the Court of Chancery, the private court of the Lord Chancellor. And both were in plain sight of the Queen's Bench, which was where trials in which the sovereign took an interest—most criminal trials, felonies, and civil wrongs touching on the security of the state—were heard.

* Probably *very* cold. In 1602, all of Europe was still experiencing the so-called Little Ice Age, during which the Thames froze over so frequently that Elizabeth I took her daily walks there.

It was an unlikely place to hear a case about a patent for playing cards.

IN ITS ORIGINAL MEANING, the word "patent" had nothing to do with the rights of an inventor and everything to do with the monarch's prerogative to grant exclusive rights to produce a particular good or service. The idea of exclusive commercial franchises crops up occasionally throughout history: Five centuries before the Common Era, the Greek colony of Sybaris granted exclusive rights for a year to a cook who invented a particularly good dish, and in the first century, a glazier supposedly presented an "unbreakable" glass cup to the Roman Emperor Tiberius, hoping for an imperial grant to manufacture it; he was disappointed when the emperor had the unfortunate soul executed (after confirming that the secret of such glass was, indeed, known only to the glazier) in order to preserve the traditional value of gold. Still, one can go centuries between occurrences. The idea started to get a bit more traction once Europe was fully embarked on the historical period known as the Renaissance; in 1421, the Lords of the Council for the city of Florence granted Filippo Brunelleschi three years' exclusive use of the boat he designed to move the stones needed to build the great Duomo.

As a *word*, "patent" enters the lexicon in something approaching its modern meaning in 1449, when the mad king Henry VI signed a document known as a *letter patent* (so called because such letters were issued openly, rather than under seal; the phrase "patently obvious" is cognate) granting a glazier named John of Utynam a twenty-year exclusive right to use his secret method for making the colored glass to be used at the chapel at Eton College. Another hundred years would pass before the next English patent, coincidentally issued to another glazier, this one named Henry Smyth, who had evidently perfected a method for making Normandy glass.*

* As distinguished from "Lorraine glass" or sheet glass, which was made from cylinders that were melted and formed into squares, Normandy glass was made from circles, or disks, and was later known as "crown glass." Aren't you glad you asked?

That patent was the first to be granted by the Tudors, but far from the last, as patents were a reliable source of revenue for a monarch whose taxing authority was severely circumscribed by Parliament, and a powerful tool for rewarding friends and promoting commerce, even to the point of encouraging skilled craftsmen to immigrate to England. By the time of the last of the Tudors—Elizabeth—the royal trade in patents was, however, dangerously out of control. She granted monopolies for the selling of salt, or making of paper, to courtiers who had two things far more important to the queen than inventiveness: loyalty and ready cash. In 1598 she issued a letter patent to Edward Darcy, a courtier ranking high enough in the Queen's regard that she admitted him to her Privy Chamber, granting him a monopoly on the manufacture, importation, and distribution of playing cards in England, evidently out of some queenly feeling that her subjects ought to be doing something better with their idle hands than dealing pasteboards with them. Unwilling to ban the practice (a slightly different monopoly had been granted by her father twenty years earlier), she was determined to regulate it, and to enrich one of her court favorites at the same time.

Unfortunately for him, three years after receiving his monopoly, in the forty-third year of her reign, Elizabeth agreed to allow her grants to be tested in the common law courts. Within months, lawyers were preparing suits intended to break one or another of these monopolies; in 1602, a competing merchant named Thomas Allein imported his own cards, and Darcy sued. In a slightly perverse reminder that lawyers have clients, not opinions, Edward Coke, as the Attorney General of England, represented Darcy, whose hostility to monopolies was already well known, though less as a matter of principle and more as a matter of economics: Coke was convinced that monopolies were costly to Britain's artisans.

In 1961, the British economist Ronald Coase published an article entitled "The Problem of Social Cost" that jump-started one of the most influential ideas in modern legal theory: the school familiarly known as Law and Economics, which proposes that legal decisions ought to account for economic efficiency as well as more traditional measures such as legislative history or case precedent. Had Coase lived three centuries earlier, he would have found Coke a most congenial colleague, since Coke's arguments against monopolies were almost entirely derived from their economic impact, specifically on the need for full employment of England's skilled craftsmen; decades before *Darcy v. Allein,* he supported the 1563 Statute of Artificers, which regulated entry into dozens of skilled crafts, set training requirements, and even allowed justices of the peace to set wages—all provisions strongly supported by the artisan guilds.

With Darcy as his client, however, Coke was trapped between his politics and his profession, and he twisted himself into a pretzel trying to reconcile the two. In order to find a precedent that confirmed his aversion to monopolies in general while still advocating on behalf of Darcy, Coke cited every obscure reference he could find, from the Bible to the Magna Carta. Revealingly, in his argument he contended that the justification for granting to the perfumed and periwigged court favorite "the franchises and privileges which the subjects have of the gift of the king" was that it was the same right that permitted an artisan to practice the special skills of his trade. Take away Darcy's right to import playing cards, the logic went, and you take away the right of a tinsmith to sell pots.

Chief Justice Popham wasn't having any. He ruled that Darcy's grant was forbidden on several grounds, all of which violated the common law. The most important one—the logic that started the ball rolling downhill toward engine 42B and *Rocket*—was the judgment of the court that the Crown could not grant a patent for the private benefit of a single individual who had shown no ability to improve the "mechanical trade of making cards," because by doing so it barred those who did. In other words, the court recognized that the nation could not grant an exclusive franchise to an individual unless that individual had demonstrated some superior "mastery" of a particular trade. Though it would be twenty years before it would be written, one of the foundations of Britain's first patent law—the doctrine that patent protection must be earned by demonstrating mastery of the method for which protection was asked—was laid.

One can imagine Coke, after this rare loss in court, exhaling with relief. The most durable constants in Coke's fifty years of legal practice were, first, his support for England's artisans, even over commercial, manufacturing, and trading interests, and second, his hostility to monopolies. It was scarcely surprising, therefore, that Coke, who had in the intervening years been made Lord Chief Justice of England, drafted the 1623 "Act concerning Monopolies and Dispensations with penall Lawes and the Forfeyture thereof," or, as it has become known, the Statute on Monopolies. The Act was designed to promote the interests of artisans, and eliminate all traces of monopolies.

With a single, and critical, exception. Section 6 of the Statute, which forbade every other form of monopoly, carved out one area in which an exclusive franchise could still be granted: Patents could still be awarded to the person who introduced the invention to the realm—to the "first and true inventor."

This was a very big deal indeed, though not because it represented the first time inventors received patents. The Venetian Republic was offering some form of patent protection by 1471, and in 1593, the Netherlands' States-General awarded a patent to Mathys Siverts, for a new (and unnamed) navigational instrument.* And, of course, Englishmen like John of Utynam had been receiving patents for inventions ever since Henry VI. The difference between Coke's statute and the customs in place before and elsewhere is that it was a *law*, with all that implied for its durability and its enforceability. Once only inventors could receive patents, the world started to change.

England's first modern patent regulation contained a number of other relevant properties:

1. The term of the patent was not to exceed fourteen years, a figure that makes sense only in terms of the artisans

* For more about the Netherlands, see chapter 11.

for whom Coke was so solicitous. Since the traditional seventeenth-century apprenticeship lasted seven years, a term of fourteen years would allow at least two cycles of apprentices to have been trained in the new industry, and therefore a generation of artisans to demonstrate their mastery of the new art.

2. The patent "must be of such manufactures, which any other at the making of such Letters Patents did not use." That is, no patent could be granted if the same process was already in use. Here the justification was political, since England's existing manufacturers and traders were utterly terrified of monopoly grants over uses already in existence. The later concepts that define patent law—that a patentable invention must be both novel and useful—make their first appearance in this portion of the Statute.
3. No patent may be "contrary to law." According to a case decided in 1572, in Coke's view, "not contrary to law" meant merely that no patent may be granted for an improvement in an existing manufacture. This happened to be the view held in England well into the eighteenth century.
4. It must not be "mischievous to the State" by raising of prices of commodities at home. Coke was of the view that the introduction of the new industry should not be granted patent protection if it resulted in a price increase, hurt trade, or was "generally inconvenient." What this meant in practice was that the Statute embodied less a love of *competition* than of a certain class of *competitors,* specifically including printers and makers of gunpowder,* reflecting his concern for employment of England's craftsmen above all.

* Coke was neither the first, nor the last, to accept "national security" exceptions to his principles.

The Statute became law in 1624. The immediate impact was barely noticeable, like a pebble rolling down a gradual slope at the top of a snow-covered mountain. For decades, fewer than six patents were awarded annually, though still more in Britain than anywhere else. It was seventy-five years after the Statute was first drafted, on Monday, July 25, 1698, before an anonymous clerk in the employ of the Great Seal Patent Office on Southampton Row, three blocks from the present-day site of the British Museum, granted patent number 356: Thomas Savery's "new Invention for Raiseing of Water and occasioning Motion to all Sorts of Mill Work by the Impellent Force of Fire."

Both the case law and the legislation under which the application was granted had been written by Edward Coke. Both were imperfect, as indeed was Savery's own engine. The law was vague enough (and Savery's grant wide-ranging enough; it essentially covered *all* ways for "Raiseing of Water" by fire) that Thomas Newcomen was compelled to form a partnership with a man whose machine scarcely resembled his own. But it is not too much to claim that Coke's pen had as decisive an impact on the evolution of steam power as any of Newcomen's tools. Though he spent most of his life as something of a sycophant to Elizabeth and James, Coke's philosophical and temperamental affinity for ordinary Englishmen, particularly the nation's artisans, compelled him to act, time and again, in their interests even when, as with his advocacy of the 1628 Petition of Right (an inspiration for the U.S. Bill of Rights) it landed him in the King's prisons. He became the greatest advocate for England's craftsmen, secure in the belief that they, not her landed gentry or her merchants, were the nation's source of prosperity. By understanding that it was England's duty, and—perhaps even more important—in England's interest, to promote the creative labors of her creative laborers, he anticipated an economic philosophy far more modern than he probably understood, and if he grew rich in the service of the nation, he also, with his creation of the world's first durable patent law, returned the favor.

Coke's motivation was not, needless to say, a longing to see steam engines decorating the English countryside, but rather a desire to see it filled with English craftsmen. A high level of craftsmanship alone, however, wasn't going to result in anything like Newcomen's engine, much less *Rocket;* artisans can be—frequently are—ingenious without being innovative. Craftsmanship needed to be married to a new way of thinking, one not yet known as the "scientific" method.

Luckily for history, a culture of observation, experimentation, and innovation was being cultivated in England at exactly the same moment that Coke was advocating for her artisans. Luckily for historians, its patron saint was not only Coke's contemporary, but his professional, political, and even romantic rival.

THE MAN WHO DIED in April 1626 with the titles Baron Verulam and Viscount St. Alban was born sixty-five years before with no title at all. Though his name at birth—Francis Bacon—was decidedly less grand, it would be recollected* in dozens of biographies and writing on everything from the birth of empiricism to the process of inductive logic. His father, Sir Nicholas, was a high-ranking member of Elizabeth's court, the Keeper of the Great Seal. His mother, Anne, was not only one of the best-educated women in England, fluent in four languages and a translator of serious Christian scholarship into English, but also the sister-in-law of William Cecil, England's de facto prime minister. It was just the sort of family that would make schooling a priority, and the beneficiary was Francis, who started a fairly traditional medieval education at Trinity College, Cambridge, and continued at the University of Poitiers, the second-oldest university in France and the alma mater of René Descartes. Poitiers was the beginning of Bacon's French connection, an affinity that shines a light on any number of subsequent events, not least his contrast with his rival Edward Coke's almost obsessive (there is no other word) *Englishness.*

* Some of that recollection includes theories that can charitably be said to be on the fringe. It is impossible to write about Bacon without mentioning Rosicrucianism, Freemasonry, and the authorship of Shakespeare's plays. Consider them mentioned.

The rivalry that features in every biography of either man probably started the day that Bacon entered Gray's Inn, another of the four Inns of Court, on June 27, 1576, and began his career in the law. And while it was not inevitable that his professional goals would result in competition with the most ambitious and successful lawyer in England—unlike Coke, Bacon would have equally prominent careers as a diplomat and philosopher—it was likely, since at that moment in history, real success depended on securing the favor of the monarch, and the rise of one courtier almost always came at the expense of another. The competition was exacerbated by what seemed to Coke an unseemly attraction to the European continent. The younger, more elegant, and more charming Bacon not only had been educated in France; almost immediately after entering Gray's Inn, he left England on a diplomatic mission, spending three years in France, Italy, and Spain, where, in rumor at least, he had a love affair with Marguerite de Valois, daughter of Henry II and Catherine de Medici. Less romantically, but more significantly, his years traveling in Europe, particularly in France and Italy, exposed him to a legal philosophy very different from the one he absorbed in London. And he liked it.

Almost all modern legal practice is derived from one of two distinct traditions. The first, the so-called civil law tradition, is a direct successor to the jurisprudence of the Roman Empire, and it dominates most of the legal systems of continental Europe; the second is the institution known as the common law, used in Britain and its former colonies. The divergence between the two dates from the eleventh century, when the only surviving copy of the complete reworking of Roman law known as the Codex Justinianus was discovered in an Italian monastery and percolated throughout Europe. The reasons for its diffusion are complicated; partly it was that it resolved an awful lot of messy contradictions between church law, local precedent, and even traditions dating back to the Visigoths. But while the kings who implemented the Justinianic Code may have liked its coherence, they adored its central theme: in Latin, *quod principi placuit, legis habet vigorem*, "the will of the prince has the force of law."

Accident as well as geography probably ensured that the idea didn't find the ground as fertile on the British side of the English Channel, where the common law continued to evolve incrementally, even haphazardly, obedient to custom rather than a grand design. It's a simplification, though not a gross one, to say that the common law is more messily pragmatic, the civil law more theoretical and pure; these are also designations that might apply, respectively, to Edward Coke and Francis Bacon.

So while Coke would use the civil law, or even sketchier precedents, when needed—he even went so far as to cite it in *Darcy v. Allein*—he remained the era's most eloquent champion of the common law, its great adversary. As Coke put it, under the common law, every man's house is his castle, not because it is defended by moats or walls, but because while the rain can enter, the king may not; under the civil law, the king is bound by nothing at all. Which was, as we shall see, a clue to the puzzle of Britain's head start in industrialization, as well as one of the sources of friction between Coke and Bacon.

It wasn't the only one. By the 1590s, and despite Bacon's connections through family and friends—he had entered into the circle surrounding the charismatic Robert Devereaux, second Earl of Essex, and another of Elizabeth's favorites—his legal career was not what one would call brilliant. In 1594, he was rejected in his bid for the position of Attorney General when the Queen selected Coke, and then, insult to injury, for Coke's former job as Solicitor General* in favor of Thomas Fleming—Coke's choice. Four years later, the always-short-of-funds Bacon tried to mend his finances matrimonially, and he set his eye on one of the wealthiest, and certainly the most fascinating, women in England. Lady Elizabeth Hatton was beautiful, witty, rich, and, with the death of her first husband in 1597, eligible. To the brilliant, handsome, but impecunious Bacon, who had known her since they were children, she was perfect. He even enlisted Essex in the service of his courtship,

* Despite the title, England's Solicitor General is almost always a barrister, not a solicitor. The position is really that of the Attorney General's chief lieutenant.

and the Earl obliged with a letter to Lady Hatton's father that read in part, "I had rather match [Elizabeth] with Mr. Bacon than with men of far greater titles." Unfortunately for Bacon, the lady had other ideas.

The clever reader will see this one coming. Coke's first wife, Bridget, with whom he was apparently deeply in love, died in June 1598. In what seemed even at the time to be the sort of haste that in later eras would have involved a shotgun, Coke and Elizabeth Hatton were married that September. Though Lady Hatton's beauty and wealth were almost certainly decisive, it is tempting to think that one of the reasons Coke pursued her so avidly was to frustrate his rival. If so, he paid for his momentary triumph with more than thirty years of grief, which included notoriously brutal domestic quarrels, vitriolic public fights, and property disputes so intractable, containing accusations of physical abuse and theft so scandalous, that they occupied England's Privy Council for three months. To call Coke's second marriage the worst decision he ever made is to understate the case; upon his death, Lady Hatton wrote "we shall never see his like again, praises be to God." She was not alone in her distaste for her husband. In a famous argument in the chambers of the exchequer in 1601, shortly after Coke had led the brutal prosecution of his onetime mentor, Essex, Bacon told his rival, "Mr. Attorney: I respect you; I fear you not, and the less you speak of your own greatness, the more I will think of it."

Coke and Bacon were so hostile to each other that they would have reflexively taken opposite sides on a debate over the best way to prepare roast beef, but for the history of the Industrial Revolution, the most significant point of divergence between the two was the way they approached the idea of innovation. One of the most consistent themes in the life of Coke, the pragmatic champion of institutions designed to protect and promote the English artisan, was his belief that the state enjoyed the most prosperity when it encouraged those artisans to perfect their craft (or at least did nothing to discourage them). Bacon, on the other hand, had an equally powerful belief in the practice of empirical science, and he advocated for a state-supported foundation that would nurture it.

That advocacy began early in Bacon's life with a powerful attack on the medieval philosophy that had dominated scholarship for eleven centuries, subordinating Greek philosophy—particularly Aristotle—to the revealed wisdom of the Bible and the Church fathers. And not just philosophy: When natural phenomena conflicted with supernatural explanation, Christian faith demanded choosing the latter, which was clearly no way to advance an understanding of the natural world. However, since people in tenth-century Europe were just as smart as they had been a thousand years before (and would be a thousand years later), their brains had to exercise themselves on *something,* and the result, from the fifth century through the fourteenth, was a network of Christian schools, and later universities, that established a classical curriculum emphasizing dialectic rather than experiment. Their leaders, known as *scholastici,* or Scholastics, filtered their Aristotle through a thoroughly Christianized screen that discarded the Greek philosopher's empirical bent and kept only his appeal to formal logic.

This was the edifice that Bacon set out to undermine when he published *The Advancement of Learning* in 1605, setting out his ambitious ideas for the reformation of education and philosophy. So ambitious were they that even listing Bacon's ideas tends to scant the depth of the footprint he left on the modern world. In fact, attempting to encapsulate his thought in a few lines or sentences is not only daunting, but insulting. Suffice it to say that before Bacon, the gold standard in inquiry was essentially contemplative and syllogistic; truth could be discovered by comparing opposing ideas. Afterward—particularly after he wrote *Novum Organum,* in 1622, in which he famously stated his belief that the compass, the printing press, and gunpowder had changed history more than any empire or religion—truth was something extracted from nature using the tools of observation and experiment. He didn't, as is sometimes suggested, invent the scientific method; he had too feeble a handle on hypotheses, and especially mathematics, to do so. But what he did understand about the scientific enterprise was profoundly important for the wave of inventions that would inundate the world a century after his death. He knew that to be self-sustaining, both science and invention needed to be *social* enterprises, depending utterly on the free flow of information among investigators. And since the state is the beneficiary of such a self-sustaining cycle of inventions, it was therefore incumbent on the state to make the information—about discoveries, techniques, and inventions—flow.

By 1608, he was making notes for the idea of a "college for Inventors [with] a Library, Vaults, fornaces [*sic*], Tarraces for Isolation, woork houses of all sorts" that could institutionalize the idea of invention in service of the state. Those notes would eventually find their way into the the posthumous publication of *The New Atlantis* in 1626, one of the achievements that appears in even the briefest biographies of the man. Bacon's fictional Atlantis was the utopian kingdom of Bensalem, located somewhere in the South Seas and distinguished by (among other things) what can only be described as a government-funded Research & Development facility. The "College of Six Days Works," or Salomon's House, was home to hundreds of investigators into the mysteries of nature; the thirty-six most senior researchers were organized by the nature of their work. Those who performed new experiments, for example, Bacon called Pioneers, or Miners, while the ones who attempted to find applications for new discoveries he called Dowry-men or Benefactors. Tellingly, though Bacon had respect for artisans and craftsmen, they were prohibited from full participation in Salomon's House. Instead, the highest honors were reserved for those "Dowry-men" who produced innovations that were of the highest value to the sovereign. In Bensalem, statues were automatically erected to honor inventors, with one of the highest to "the inventor of ordnance and of gunpowder." Salomon's House was the explicit inspiration for the Royal Society, which was founded twenty-four years after Bacon's death in large part to honor his notion of science as a collaborative venture whose proper goal is the material improvement of mankind.

The material improvement of mankind, not of inventors. In Bensalem, inventors were not granted any property rights in their inventions: no patents. In the conflict over the Statute on Monopolies, as with much else, Bacon stood on the opposite side from Coke, and his ideal society, not at all surprisingly, reflected his belief that the surest route to innovation was by relying on men who wanted only fame as a reward. Bacon's faith in progress through collective action by public servants obliged him to reward innovators; his snobbery made it impossible to make those rewards commercially valuable, particularly to anyone who might need the income, and *especially* in the form of ownership of inventions.

Coke's pragmatic support for the artisan class and Bacon's vision of invention as a collective endeavor were both, for their day, progressive—both were willing to grant the highest honors to any accomplished man of low birth—but nonetheless a bit nearsighted: Both could see the value of the innovator to *society,* but misunderstood how to align social benefits with individual rewards; and Bacon, in particular, was unwilling to grant inventors even a temporary property right in their ideas.

That sort of ownership required a revolution in the idea of property itself.

ON TUESDAY, FEBRUARY 12, 1689, the royal yacht *Isabella* docked at Gravesend carrying Mary Stuart, the daughter of King James II, on her last day as the Princess of Orange. A day later, in London, she would be acclaimed Mary II, Queen of England, thus putting an end to the most tempestuous four decades in English history. Any fifty-year-old Englishman at Gravesend that day had lived through two or three civil wars; the execution of one Charles and the exile of another; the autocratic rule of Oliver Cromwell; the restoration of Charles II; the 1665 recrudescence of the bubonic plague; and the Great Fire of London the following year. The fifty-year-old would be able to recall more recent events as well, including the rebellion of the Duke of Monmouth and any number of near rebellions beginning with the 1685 death of Charles and the accession of his younger brother, James, whose conversion to Catholicism failed to reignite civil war only because his Protestant daughter, the wife of the Dutch stadtholder William of Orange, was the heiress presumptive.

So when the *Isabella* arrived with that same daughter, a few months after her husband had successfully asserted her rights at the head of an invading army, she was conveying a much-longed-for respite from what, in the apocryphal curse, are known as "interesting times." And that wasn't all. The royal yacht led a flotilla that carried, as a passenger, a sometime poet, essayist, and physician, a fifty-six-year-old man who had been living as an exile in the Dutch cities of Amsterdam, Leiden, and Utrecht for the preceding four and a half years, using the *nom de refuge* of "Dr. Van der Linden." His real name was John Locke, and he was, in the words of Thomas Jefferson, one of the three greatest men who ever lived (the other two were Isaac Newton—and Francis Bacon).

Son of a lawyer, grandson of a clothier, Locke had successively been a brilliant but bored student at Westminster School and at Christ Church College, Oxford, where his fellow students included John Dryden, Robert Hooke, and Christopher Wren. It is tempting to think that he cultivated the company of men who formed the kernel of the future Royal Society when they were all at Oxford, but if so, it has escaped the attention of hundreds of biographers.

Locke's Oxford career gave only hints of his future prominence. Despite his steady climb up the academic ladder, starting as a "student"—at the time, a formal title more like "fellow"—by 1660 becoming a lecturer and reader in rhetoric, and by 1664 the "censor of moral philosophy," he wasn't what one might call an academic star. In the words of Lady Damaris Masham,* he "had so small satisfaction from his studies" that he failed to work very hard at them. Whether out of boredom or out of resistance to the pressures to take instruction as an Anglican clergyman, he subsequently dab-

* The remarkable Lady Masham was not simply Locke's first biographer and friend but the first Englishwoman to publish philosophical writings on her own account, and a regular correspondent with, among others, Gottfried Wilhelm Leibniz.

bled in law, medicine, diplomacy, and natural philosophy. In the last capacity, he collaborated with Robert Boyle, who sponsored him for a fellowship in the Royal Society, though Locke's involvement was scarcely life-changing; in his four decades as a member of the RS, his only publication was a letter regarding a poisonous fish sent to him by a friend in the Caribbean.

What *did* change Locke's life in matters large and small was his relationship with Anthony Ashley-Cooper—Lord Ashley when Locke first met him, and later the first Earl of Shaftesbury, the Lord High Chancellor of England, President of the Privy Council, and ultimately the great political adversary of Charles II. It was in service to Shaftesbury, originally as a physician, that Locke developed temperaments heretofore absent, including a toleration of nonconformist ideas and, more relevant, an interest in the economic relations between men. In 1668 he even wrote a treatise with the wonkish title *Some of the Consequences that are Like to Follow upon Lessening of Interest to 4 per cent.* And it was in service to Shaftesbury that Locke was forced to leave England not once, but twice. The first time he exiled himself as a precaution, after Shaftesbury's circle published a series of arguably seditious pamphlets; when the furor had died down, in 1679, he returned. Four years later, however, after the group planned, though never carried out, an assassination attempt on King Charles, Locke, now an accused traitor and fugitive, was forced to escape again, this time to Rotterdam, where he arrived on November 1, 1683, there to stay until he could return more than five years later. With him when he arrived on the *Isabella* that February day in 1689 was the draft manuscript for a work that would be published later that year under the title *Two Treatises on Government.**

* Or probably the manuscript. From internal evidence—references to "King James" rather than "James II" suggest that at least portions of the *Treatises* were written before James II's accession in 1685—it seems safe to assume that tucked away in the one-time exile's luggage were the draft treatises. Scholars still debate whether they were written in the heat of the controversy over the Exclusion Bill, a statute introduced by Shaftesbury to exclude the now publicly Catholic James II from the throne. If so, they are a powerful argument that engagement in the rough-and-tumble of political life is no barrier to producing original and hugely influential political philosophy.

As with Bacon, there is no way that Locke's thoughts on the nature of knowledge, religious freedom, political organization, or a dozen other subjects can be contained within a single book, much less a portion of a single chapter. Those wondering what on earth Locke's writings have to do with the world's first steam locomotive, however, will find that the relevant passages from the *Treatises* concern Locke's views on the rather slippery idea of property: or, rather, the idea that ideas *are* property.

Despite much evidence to the contrary, the first word most children learn (after the local equivalent of "mama" and "papa") is not "mine." Nonetheless, the concept of ownership in some form is pretty universal, even in eras and cultures that deny it. This is because the idea of property is essential for a culture to be able to speak of the relationship between people and things. And some sort of property law is utterly necessary if a society is to resolve disputes between people *over* things. The difference between the way property is understood on Wall Street and in an Egyptian souk clearly shows that the concept is absolutely contingent upon culture, but in the societies that trace their legal and philosophical systems back to Greece and Rome, property traditionally has three characteristics, none of them absolute:

1. Exclusive possession (which also obliges everyone else to keep away);
2. Exclusive use; and
3. Some right of conveyance, the ability to transfer the property right.

The Western idea of property is distinctive not because of these three characteristics, but because of its tendency to generalize them; in many non-Western societies, the rules for land may have little or nothing to do with the rules for animals, which in turn may be unrelated to the rules for objects. In the Western tradition, the default position is that individual possession/use/conveyance is the zero point from which exceptions may deviate. This perspective may have begun as nothing more than a convenience—it is easier to adjudicate disputes if all the rights in question are decided together—but like a ratchet that turns in only one direction, it has tended, over the centuries, to promote the accumulation of more and more individual rights over things, such as land, and even water, that were once held in common.

That "accumulation," of course, was generally done at the point of a sword, which seemed to Locke to be a historical *fact*, but not a historical *right*. And rights were what interested Locke. His development of a new definition of property rights evolved over decades, and it owed as much to the recent history of Britain as it did to the Roman idea of *usufructus* (the right to profit from a thing without owning it) or the medieval notion of *seisin* (possession of land without title to it). The Civil War, in particular, had turned the world upside down in more ways than one; there was something dangerously explosive about gathering a large number of people by persuading them that God wants them to be free of both political and religious tyranny, and then giving them weapons. Four years after the start of hostilities, a mass movement within Oliver Cromwell's New Model Army (the first in history to give its soldiers a say in their own governance, an idea that seems as radical today as it did to Cromwell) demanded the right to vote in their "Agreement of the People" of 1647. The conflict prompted a public debate, at Putney in southwest London, between the self-named "Levellers," the element among the Parliamentarians most in favor of democratic reforms as part of the battle against the royalists, and the Army's more conservative leadership, and the discussion moved inevitably from voting rights to property rights, since the former depended on the latter.

Common law tradition granted the franchise only to owners of either a specified amount of land or of a license to trade, under the logic that since laws are made to safeguard property, legislators should be elected only by those with an interest in those laws, or, as the conservative faction had it, "disposing of the affairs of the kingdom [requires] a permanent fixed interest in this kingdom." The Leveller response was that every man—almost every man; they excluded servants and beggars, for example—has property in his own person and has therefore an interest in parliamentary action.

If the Levellers were the radicalized members of the Parliamentary party, the Diggers, an agrarian communist movement that emerged in 1649, were the radical edge of the Levellers. Their concept of legitimate property was far different, as were their demands, including the right to common land: a return to the state of nature before the appropriation of that land by others. Gerard Winstanley, leader of the Diggers, a onetime cloth merchant who lost his livelihood, and then his family's land, as a side effect of the Civil War, is remembered as being violently opposed to the very idea of property. In the Digger manifesto, a "Declaration from the Poor Oppressed of England," he pulled no punches, telling the gentry, "You and your ancestors got your propriety [i.e. property] by murder and theft, and you keep it by the same power from us." But his understanding of property was almost completely limited to that special form known as real estate. This made sense, of course; at a time when virtually everything of value, in Britain and everywhere, was either land or the produce of land, property and land were functionally synonymous. And since the amount of land was essentially fixed, it could be possessed by one man only if he dispossessed another.

Enter John Locke, whose central premise was that man has no right to own the work of God—to own land—but that rightful property is derived from the labor of man mixed with that of God. That is, when man combines his labor with the goods of the earth, he has created a natural right to the product. The right predates government, law, or kings, and is therefore present in his hypothetical state of nature. By deriving the right from the biblical grant by God to Adam of the earth for his subsistence, Locke reasoned his way to the idea that the earth is no good to any particular man unless someone labors to make it so. Locke thus triangulated between the democratic Levellers (and the Diggers who shared an enthusiasm for inventors and inventing: Winstanley wrote, in 1652, "Let no young wit be crushed in his invention. . . . Let every one who finds out an invention have a deserved honour given him") and the status quo, arguing that the then current division between haves and have-nots was legitimate so long as the cause of the division was labor.

THE PATIENT READER IS now asking, "What does this have to do with steam power?" (The impatient ones asked it twenty pages ago.) This: By equating labor with a property right, Locke found a right to property *anywhere labor is added.* The defining characteristic became the labor, not the thing. And labor, in Locke's formulation, was as much of mind as of muscle. "Nature furnishes us only with the material, for the most part rough, and unfitted to our use; it requires labour, art, and thought, to suit them to our occasions. . . . Here, then, is a large field for knowledge, proper for the use and advantage of men in this world; viz. to find out new inventions of despatch to shorten or ease our labour, or applying sagaciously together several agents and materials, to procure new and beneficial productions fit for our use, whereby our stock of riches (i.e. things useful for the conveniences of our life) may be increased, or better preserved: and for such discoveries as these the mind of man is well fitted."

So, while Edward Coke's Statute on Monopolies established England's first patent law, the general acceptance of the notion of what we would now call intellectual property awaited its articulation by John Locke.* It is scarcely surprising that the Copyright Law of 1710 appeared so soon after Locke's works, followed by the 1735 Engraver's Act, which granted the same rights to prints as the Copyright Act did to literary works.

This does not mean that Locke's ideas swept all earlier ones away, any more than the Statute on Monopolies caused an immediate explosion in patent grants. Ideas, and the institutions that promote them, take some time to take root. Locke's own protégé,

* Three hundred years later, a group of mathematically minded economists would distinguish between tangible and intellectual property in much the same way, as we shall see in chapter 11.

David Hume, was never persuaded that property rights derived from natural law. Eighty years after Locke's death, conservatives like Edmund Burke, and progressives like Jeremy Bentham and John Stuart Mill, were still uncomfortable with Locke's idea of natural laws; Bentham called them "nonsense on stilts." The final victory, however, was Locke's; in 1776, Adam Smith was virtually channeling Locke's *Second Treatise,* writing in *The Wealth of Nations,* "The property which every man has in his own labour, as it is the original foundation of all other property, so it is the most sacred and inviolable." Smith's French counterpart, Anne-Robert-Jacques Turgot, echoed him: "God . . . made the right of work the property of every individual in the world, and this property is the first, the most sacred, and the most imprescriptible of all kinds of property."

Recognition of a property right in ideas was the critical ingredient in democratizing the act of invention. However imperfectly, Coke's patent system, combined with Locke's labor theory of value, offered a protected space for inventive activity. The protected space permitted, in turn, the free flow of newly discovered knowledge: the essence of Francis Bacon's program. Once a generation of artisans discovered they could prosper from owning, even temporarily, the fruits of their mental labor, they began investing that labor where they saw the largest potential return. Most failed, of course, but that didn't stop a trickle of inventors from becoming a flood.

The reason that that flood would, eventually, find its way to engine 42B and *Rocket*—and would become a river instead of a lake—was an unprecedented fusion of theory, experiment, and measurement, which is explored in the next chapter.

CHAPTER FOUR

A VERY GREAT QUANTITY OF HEAT

concerning the discovery of fatty earth; the consequences of the deforestation of Europe; the limitations of waterpower; the experimental importance of a Scotsman's ice cube; and the search for the most valuable jewel in Britain

THE GREAT SCIENTIST AND engineer William Thomson, Lord Kelvin, made his reputation on discoveries in basic physics, electricity, and thermodynamics, but he may be remembered just as well for his talent for aphorism. Among the best known of Kelvin's quotations is the assertion that "all science is either physics or stamp collecting" (while one probably best forgotten is the confident "heavier-than-air flying machines are impossible"). But the most relevant for a history of the Industrial Revolution is this: "the steam engine has done much more for science than science has done for the steam engine."

For an aphorism to achieve immortality (at least of the sort certified by *Bartlett's Familiar Quotations*), it needs to be both true and simple, and while Kelvin's is true, it is not simple, but simplistic. The science of the eighteenth century didn't provide the first steam engines with a lot of answers, but it did have a new, and powerful, way of asking questions.

It is hard to overstate the importance of this. The revolution in the understanding of every aspect of physics and chemistry was built on a dozen different changes in the way people believed the world worked—the invariability of natural law, for example (Newton famously wrote, "as far as possible, assign the same causes [to] respiration in a man, and in a beast; the descent of stones in Europe and in America; the light of our culinary fire and of the sun") or the belief that the most reliable path to truth was empirical.

But scientific understanding didn't progress by looking for truth; it did so by looking for mistakes.

This was new. In the cartoon version of the Scientific Revolution, science made its great advances in opposition to a heavy-handed Roman Catholic Church; but an even larger obstacle to progress in the understanding and manipulation of nature was the belief that Aristotle had already figured out all of physics and had observed all that biology had to offer, or that Galen was the last word in medicine. By this standard, the real revolutionary manifesto of the day was written not by Descartes, or Galileo, but by the seventeenth-century Italian poet and physician Francesco Redi, in his *Experiments on the Generation of Insects,* who wrote (as one of a hundred examples), "Aristotle asserts that cabbages produce caterpillars daily, but I have not been able to witness this remarkable reproduction, though I have seen many eggs laid by butterflies on the cabbage-stalks. . . ." Not for nothing was the motto of the Royal Society *nullius in verba:* "on no one's word."

This obsession with proving the other guy wrong (or at least testing his conclusions) is at the heart of the experimental method that came to dominate natural philosophy in the seventeenth century.* Of course, experimentation wasn't invented in the seventeenth century; four hundred years earlier, while Aquinas was rejiggering Aristotle for a Christian world, the English monk

* The modern definition of experimentation—isolation of a single variable, to test and record the effect of changing it—still lay a hundred years in the future. We will meet the creator of this sort of experimental design, John Smeaton, in chapter 6.

Roger Bacon was inventing trial-and-error experimentation—in Europe, anyway; experimentation was widely practiced in medieval Islamic cities from Baghdad to Córdoba. Bacon was, however, a decided exception. The real lesson of medieval "science" is that the enterprise is a social one, that it was as difficult for isolated genius to sustain progress as it would be for a single family to benefit from evolution by natural selection. Moreover, even when outliers like Friar Bacon, and to a lesser degree the era's alchemists, engaged in trial-and-error tests, they rarely recorded their results (this might be the most underappreciated aspect of experimentation) and even more rarely shared them. A culture of experimentation depends on *lots* of experimenters, each one testing the work of the others, and doing so publicly. Until that happened, the interactions needed for progress were too few to ignite anything that might be called a revolution, and certainly not the boiler in *Rocket*'s engine.

It took a massive shift in perspective to create such a culture, one in which a decent fraction of the population (a) trusted their own observations more than those made by Pliny, or Avicenna, or even Aristotle, and (b) distrusted the conclusions made by their contemporaries, at least until they could replicate them. In the traditional and convenient shorthand, this occurred when "scientific revolutionaries" like Galileo, Kepler, Copernicus, and Newton started thinking of the world in purely material terms, describing the world as a sort of machine best understood by reducing it to its component parts. The real transformation, however, was epistemological: Knowledge—the same stuff that Locke was defining as a sort of property—was, for the first time in history, conditional. Answers, even when they were given by Aristotle, were not absolute. They could be replaced by new, and better, answers. But a better answer cannot be produced by logic alone; spend years debating whether the physics of Democritus or Leucippus was superior, and you'll still end up with either one or the other. A new and improved version demanded experiment.

If the new mania for scientific experimentation began sometime in the sixteenth century, with Galileo—even earlier, if you want to begin with René Descartes—it took an embarrassingly long time to contribute much in the way of real-world technological advances. Francis Bacon might have imagined colleges devoted to the material betterment of mankind, in which brilliant researchers produced wonders that might allay hunger, cure disease, or speed ships across the sea; but the technology that mostly occupied the Scientific Revolution of the sixteenth and seventeenth centuries was improving scientific instruments themselves (and their close relations, navigational instruments). Science did build better telescopes, clocks, and experimental devices like von Guericke's hemispheres, or Hooke's vacuum machine, but remarkably little in the way of useful arts. The chasm that yawned between Europe's natural philosophers and her artisan classes remained unbridged.

Describing how that bridge came to be built has been, for decades, the goal of an economic historian at Northwestern University named Joel Mokyr, who knows more than is healthy about the roots and consequences of the Industrial Revolution. In a series of books, papers, and articles, Professor Mokyr has described the existence of an intellectual passage from the Scientific Revolution of Galileo, Copernicus, and Newton to the Industrial Revolution, which he has named the "Industrial Enlightenment"—an analytical construct that is extraordinarily useful in understanding the origins of steam power.

The beauty of Mokyr's analysis is that it replaces an intuitive notion—that the Industrial Revolution must have been *somehow* dependent upon the Scientific Revolution that preceded it—with an actual mechanism: in simple terms, the evolution of a market in knowledge.

The sixteenth and seventeenth century's Scientific Revolution was a sort of market, though the currency in which transactions occurred was usually not gold but recognition: Gaspar Schott saw Otto Gericke's vacuum experiments and wrote about them; Boyle read his account and published his own. Huygens, Papin, and Hooke all published their own observations and experiments. They had an interest in doing so; as a class, they generally sought pride rather than profit for their labors, and were therefore paid with notoriety, along with some acceptable sinecure: professorships, pensions, patronage. They even sometimes, as with Hooke's attempt to turn his discovery of the Law of Elasticity into a balance spring mechanism for a marketable timepiece, showed decided commercial impulses. But the critical thing was that a structure within which scientists could trade their newly created knowledge had been evolving for nearly a century before it was widely adopted by more commercially minded users.

Their need for it, however, was enormous. Prior to the eighteenth century, innovations tended to stay where they were, since finding out about them came at a very steep price; in the language of economists, they carried high information costs. For centuries, a new and improved dyeing technique developed by an Italian chemist would not be available, at any affordable cost, to a weaver in France, both because the institutions necessary for communicating them, such as transnational organizations like the Royal Society, did not exist, and because the value of the innovation was enhanced by keeping it secret.

For a century, that was how things stood. Europe's first generation of true scientists produced a flood of testable theories about nature—universal gravitation, magnetism, circulation of the blood, the cell—and tools with which to understand them: calculus, the microscope, probability, and hundreds more. But this flood of what Mokyr calls *propositional knowledge* did not diffuse cheaply into the hands of the artisans who could put them to use, since the means of doing so depended on a sophisticated publishing industry producing books in Europe's vernacular languages rather than the Latin of scientific discourse, and on a literate population to read them.

An even bigger problem was this: as the seventeenth century wound down, scientific knowledge was becoming a public good, partly because of what we might call the Baconian program. Francis Bacon's vision of investigators and experimenters working in a common language for the common good had inspired an entire generation; and, to be fair, the extraordinary number of related discoveries in mathematics, physics, and chemistry had indeed benefited everyone. But partly it was a matter of class. Scientists in the seventeenth and eighteenth centuries, though a highly inventive bunch, were members of a fraternity that depended on allegiance to the idea of open science—so much so that even Benjamin Franklin, clearly a man with a strong commercial sense, did not as a matter of course take out patents on his inventions. The result was what happens when work is imperfectly aligned with rewards: Science remained disproportionately the activity of those with outside income. Predictably, Bacon's *New Atlantis* model, which worked so well for the diffusion of scientific innovations, had built in a limit on the population of innovators.

By the start of the eighteenth century, however, things were changing, and changing fast. Artisans like Thomas Newcomen and itinerant experimentalists like Denis Papin were both corresponding with Robert Hooke. Engineers like Thomas Savery were demonstrating inventions in front of the physicists and astronomers of the Royal Society. Most important, mechanics, artisans, and millwrights, who had been taught not only to read but to measure and calculate, started to apply the mathematical and experimental techniques of the sciences to their crafts. *Useful knowledge* (the historian Ian Inkster calls it *useful and reliable knowledge,* or URK) became, in Mokyr's words, "the buzzword of the eighteenth century."

The same mechanisms that spread the discoveries of the Scientific Revolution throughout Europe—correspondence between researchers, and publications like the Royal Society's *Philosophical Transactions*—proved just as useful in the diffusion of applied knowledge. But because Europe generally, and Britain specifically, had a lot more artisans than scientists, the demand for commercially promising applications was far greater than those with a purely scientific bent. New ways of buying and selling applied knowledge emerged to meet the demand. J. T. Desaguliers, the same critic who had sniffed at Thomas Newcomen's mathematical training, spent decades giving a hugely popular series of lectures all across rural England and later collected them in his

1724 *Course of Mechanical and Experimental Philosophy.* By the 1730s, millwrights, carpenters, and blacksmiths were able to purchase what we would today call a continuing education in pubs and coffeehouses in the craft they had learned as apprentices. By 1754, the drawing master William Shipley could found the Royal Society of Arts (at a time that made no distinction between fine, decorative, or applied arts) on a manifesto that argued, "the Riches, Honour, Strength, and Prosperity of a Nation depend in a great Measure on Knowledge and Improvement of useful Arts [and] that due Encouragement and Rewards are greatly conducive to excite a Spirit of Emulation and Industry. . . ." Britain's artisans were now buyers at their own knowledge market, and they were doing so to fatten not their reputations, but their wallets.

One of the criticisms often made of economists is that they see all of human behavior as a kind of market. But neither steam engines in general, nor *Rocket* in particular, makes much sense without referring to an entire series of markets: one for transportation of Manchester cotton, another for the iron on which the engine ran, still another for the coal it burned, and so on. The most important of all, however, was the Industrial Enlightenment's de facto market in what would one day be called "best practices" from the craft world. By the first decades of the eighteenth century, a market had emerged in which an English ironmonger could learn German forging techniques, and a surveyor could acquire the tools of descriptive geometry.

But markets do more than bring buyers and sellers together. They also reduce transaction costs. One of those costs, in the early decades of the eighteenth century, was incurred due to the fact that an awful lot of the newest bits of useful knowledge were hard to compare, one with the other, because they described the same phenomenon using different words (and different symbols). As the metaphorical shelves of the knowledge market filled with innovations, buyers demanded that they be comparable, which led directly to standardization of everything from mathematical notation to temperature scales. In this way, the Industrial Enlightenment's knowledge economy lowered the barriers to communication between the creators of theoretical models and masters of prescriptive knowledge, for which the classic example is Robert Hooke's 1703 letter to Thomas Newcomen advising him to drive his piston by means of vacuum alone.

The dominoes look something like this: A new enthusiasm for creating knowledge led to the public sharing of experimental methods and results; demand for those results built a network of communication channels among theoretical scientists; those channels eventually carried not just theoretical results but their real-world applications, which spread into the coffeehouses and inns where artisans could purchase access to the new knowledge.

Put another way, those dominoes knocked down walls between theory and practice that had stood for centuries. The emergence of a market in which knowledge could be acquired for application in the world of commerce had also increased the population capable of producing that knowledge. It would occur in the study of medicine, of chemistry, and even of mathematics, but nowhere was it more relevant to the future of industrialization than in the study of the science of heat.

TWO YEARS BEFORE HIS death in 1704, John Locke collaborated with William Grigg, the son of one of Locke's oldest friends, to produce an interlineary translation—that is, alternating lines of Latin and English—of Aesop's fables. One of those fables, "De sole et vento," or "The Sun and the Wind," famously recounts the contest between the two title characters over which could successfully cause a traveler to remove his coat. It is among the earliest, and is certainly one of the best known, accounts of the debate between heat and cold. Or, as we would call it today, thermodynamics.

Though the equations of thermodynamics are obviously essential to understanding the machine that Newcomen and Calley demonstrated in front of Dudley Castle, they were just as obviously unnecessary for building it. What the ironmonger and glazier didn't know about the physics of the relationship between water and steam would fill libraries, while what they *did* know was mostly wrong. This is in no way a criticism of the inventors; what *everyone* knew, at the time, was mostly wrong.

Even the seventeenth century's newfound affinity for experimental science hadn't done much to correct misapprehensions about the nature of heat. When Francis Bacon (to be fair, more a philosopher of science than a scientist) attempted, in 1620, an exhaustive description of the sources of heat, he included not only obvious candidates like the sun, lightning, and the "the violent percussion of flint and steel," but also vinegar, ethanol ("spirits of wine"), and even intense cold. He also failed to produce anything like a testable theory; while he did nod toward equating heat with motion, he failed to realize that heat was a measurable quantity—the first thermometers that used any sort of scale date from the early eighteenth century; imagine, if you can, drawing a map without knowing the number of inches to the mile, and you can see the obstacle this presented. Galileo, Descartes, and especially Robert Boyle also tried explain how motion was related to heat, particularly friction. They each failed, which is not surprising; nearly three centuries later, Lord Kelvin himself was still unsure whether heat energy could be equated with mechanical energy.

The reason is that seventeenth-century heat theorists were hamstrung by the two existing models from the world of natural philosophy. The first was the notion that heat was an "elastic fluid" or gas; the other, that it was a consequence of exciting the motion of an object's constituent parts, which were known as "atoms," though those who used the term didn't mean the same thing as a modern chemistry textbook. Isaac Newton had demonstrated that the best escape from the prison of Aristotelian ideas about motion was an entirely new set of invariant laws, but Newton, curious though he was, showed only small interest in the scientific nature of heat. As a result, the first really useful theory of heat and combustion was being articulated elsewhere.

In 1678, less than a decade before Newton introduced the world to the laws of motion and universal gravitation in the first edition of *Principia Mathematica* (and two decades before Savery received patent number 376), the alchemist Johann Joachim Becher departed the patchwork quilt of grand duchies, principalities, and free cities that the Thirty Years War had made of Germany. By then, he had already served as a court physician to the Elector of Bavaria; as a secret agent in the pay of the Austrian Emperor; and as a special emissary for Prince Rupert, the onetime commander of royalist cavalry during the English Civil War. It was in the last capacity that he journeyed to Scotland and Cornwall, to examine and report on the coal and tin mines of Britain. He also had a personal motive: to discuss his discovery with the new Royal Society.

Becher called the substance he had "discovered" *terra pingua,* which confusingly translates as "fatty [by which he meant *inflammable*] earth." This was a thoroughly respectable attempt to reconcile the established belief that the world was made up of the ancient four elements—fire, water, earth, and air—with the observation that the phenomenon of combustion seems to involve them all; that in some way, the process that burns wood is similar to the one that causes iron to rust, if only because the absence of air prevents both. Becher's discovery, renamed in 1718 as *phlogiston,* replaced the Aristotelian elements with a different foursome: water; *terra mercuralis,* or fluid earth (i.e. mercury, and similar substances); inert earth, or *terra lapida* (that is, salts); and Becher's *terra pingua,* thus covering all possible forms of matter, and demoting fire from an element to a phenomenon. The theory that explained the behavior of Becher's inflammable earth still has, in some circles, the flavor of charlatanism, and to be sure, Becher wasn't completely free of the taint; he had spent years trying to sell a method for turning sand into gold. However tempting it is to poke fun at the scientific ignorance of our ancestors, though, in the case of the phlogiston theory, it is a temptation that should be resisted. Though phlogiston theory is wrong, it is considerably more scientific than is generally understood, and it was an early and necessary step on the way to a proper understanding of thermodynamics, and of the way in which *Rocket* transformed heat into movement.

At the core of the theory is the idea that anything that can be burned must contain a material—phlogiston—that is released by the process of burning. Once burned, the dephlogisticated substance becomes *calx* (an example would be wood ash), while the air surrounding it, which was known to be essential to combustion, became *phlogisticated.* Thus, burning wood in a sealed chamber could never result in complete combustion, because the dephlogisticated air necessary for burning became saturated with phlogiston. The reason that wood ash weighs less than wood, therefore, is because of the loss of phlogiston to the air when it is partially burned.

However, any theory of heat transfer that depended upon the swap of a *substance* demanded that it go somewhere. Phlogiston theory worked fine for those things that weighed less after heating something else, but it was vulnerable to an encounter with any substance that didn't. Magnesium, for example, seems to gain weight when heated (actually, it becomes magnesium oxide). Heat can be transferred even when "condensed phlogiston" doesn't change at all. A red-hot hunk of iron will cause water to steam even if it weighs the same after it is cooled by that same water.

By the middle of the eighteenth century, despite some truly passionate devotees, most especially the English chemist Joseph Priestley, phlogiston theory was displaced, largely by the work of the French scientist Antoine Lavoisier. Which is why phlogiston theory deserves a bit more respect than it is generally given. It is a goofy theory, to be sure, with funny-sounding names for its fundamental concepts (though no funnier-sounding than quarks, Higgs bosons, or other notions from the world of quantum physics). But it *is* a theory, in a way that the four elements of antiquity were not. Phlogiston was incorrect in its particulars: The relationship between fire and rust is that both are examples of what happens when oxygen, which would not be discovered for another century, reacts with another substance. But it was also testable, in the sense made famous by the philosopher of science Karl Popper. Phlogiston theory could be proved false, and eventually was. The first to do so was a pioneering physicist and chemist at the University of Glasgow, a key figure in the evolution of the steam engine, named Joseph Black.

BLACK WAS A THOROUGHGOING Scot, despite his Bordeaux birthplace, an incidental consequence of his family's involvement in the wine trade, and his early schooling in Belfast. He matriculated at both Glasgow and Edinburgh universities, and subsequently served as professor of chemistry at first one and then the other, ending up at Edinburgh in 1766. Long before that, he had demonstrated a remarkable gift for experimental design, and what was, for the time, painstaking care in experimentation itself, particularly into the nature of heat.

The gift for designing experiments was much on display in Black's research into the nature of what a later science of chemistry knows as carbon dioxide. He was not, by all accounts, much interested in testing phlogiston theory when he began; instead, as a physician, he was looking for a way to dissolve kidney stones. His investigations accordingly began with an investigation into the then well known process by which chalk, or calcium carbonate, turns into the caustic quicklime, which was the name then used for calcium oxide. Black chose to work with a similar substance: magnesium carbonate, then known as *magnesia alba.* Since the transformation required combustion, at very high heat, phlogiston theory suggested that the reason was the absorption of the fiery substance by the chalk. Black, by careful experiment, showed that the magnesia alba weighed less after heating, but regained precisely the same amount when cooled in the presence of potash, from which he reasoned that the substance that departed the original substance—CO_2—had returned. He did not, of course, put it quite that way, since oxygen itself still awaited discovery some decades later. Instead, he wrote, "Quick-lime [i.e. calcium oxide] therefore does not attract air when in its most ordinary form, but is capable of being joined to one particular species only, which is dispersed thro' the atmosphere, either in the shape of an exceedingly subtle powder, or more probably in that of an elastic fluid [which I have called] 'fixed air.'" Or, as your high school chemistry teacher would explain it, calcium oxide becomes calcium carbonate in the presence of carbon dioxide.

This discovery alone, which was the first test that phlogiston theory failed, would have purchased for Professor Black a place in the history of science. But what earned him a place in the story of steam power were his subsequent experiments on the nature of heat itself. Or, more accurately, on the nature of ice.

Water, as we have seen, is a most curious substance: In both its gaseous and solid states, it occupies more volume than it does as a liquid. It is also (practically uniquely) present on earth as a solid, a liquid, and a gas. By 1760, Black had become fascinated by the properties of water in its solid version, and even more fascinated by the transition from one phase to another. He was particularly intrigued by the fact that frozen water, whether in the form of ice or snow, did not melt immediately upon coming into contact with high heat, but did so gradually. For another curious fact is that a glass with ice in it will stay the same temperature—a little above 32°F or 0°C—whether it has six unmelted ice cubes in it or one. The temperature starts to rise only when *all* the ice is melted. In the same way, a pot of water brought to a boil does not thereafter increase in temperature, no matter how hot the fire underneath it. These are by no means intuitive results, but Black observed them again and again, once again finding phlogiston theory insufficient to explain the phenomena. Instead, he came up with an idea of his own, called *latent heat,* which he defined as the amount of heat gained or lost by a particular substance before it changes from one physical state to another—gas to liquid, solid to liquid. To Black, latent heat was the best way to explain the fact that water, when it nears its boiling point, does not suddenly turn to steam with, in his words, "a violence equal to that of gunpowder."

The experiments that confirmed this hypothesis were simple, and ingenious. Black took a quantity of water and, using a thermometer, took its temperature. He then placed the water over heat and measured both the amount of time it took for the water to boil and the amount of time it took, once boiling, to boil away completely. By comparing the two, he established the amount of heat the water continued to absorb after its own temperature stopped rising. Many years later, Black described his discovery:

> I, therefore, set seriously about making experiments, conformable to the suspicion that I entertained concerning the boiling of fluids. My conjecture, when put into form, was to this purpose. I imagined that, during the boiling, heat is absorbed by the water, and enters into the composition of the vapour produced from it, in the same manner as it is absorbed by ice in melting, and enters into the composition of the produced water. And, as the ostensible effect of the heat, in this last case, consists, not in warming the surrounding bodies, but in rendering the ice fluid; so, in the case of boiling, the heat absorbed does not warm surrounding bodies, but converts the water into vapour. In both cases, considered as the cause of warmth, we do not perceive its presence: it is concealed, or latent, and I give it the name of LATENT HEAT . . .

Thus, Black calculated that a pound of liquid water had a latent heat of *vaporization* of 960°F; its latent heat of *fusion*—the amount of heat ice absorbs before completely melting—he measured at 140°F.* That is a *lot* of latent heat. Water absorbs nearly three times the amount of heat before vaporizing as the same quantity of ethanol, one of the many reasons that your waiter can flambé brandy, but not orange juice. Again, Black's experimental and quantitative mind used a different sort of arithmetic: He heated a pound of gold to 190° and placed it in a pound of water at a temperature of 50°; when he took the temperature of the

* Modern engineers generally measure this as kilojoules/kilogram, but in British Thermal Units (the amount of heat needed to raise the temperature of a pound of water 1°F) the numbers are 144 and 965 respectively. Thus, it takes 144 BTUs to turn a pound of ice at 32° into a pound of water at 32°, and 965 BTUs to convert a pound of water at 212° into steam.

combined elements and found it to be only 55°, he concluded that water had nearly twenty times more capacity for heat than did gold.

It took a pretty big fire, therefore, to boil the water in the atmospheric engine Thomas Newcomen had erected in front of Dudley Castle that day in 1712. Joseph Black had discovered a new way of measuring how big, but the relevant metric for Newcomen and Calley wasn't degrees Fahrenheit. It was fuel.

This simple fact was, in its way, as revolutionary as Coke's Statute or Newton's Laws of Motion. For millennia, advances in the design of machines to do work had been driven entirely by measures of their output: a tool that plows more furrows, or spins more wool, or even pumps more water, was ipso facto a better machine. Prior to the seventeenth century, the choices for performing such work—defined, as it would be in an introductory physics class, as the transfer of energy by means of a force—had been made from the following menu:

- Muscle, either human or animal;
- Water; or
- Wind.

Muscle power is, needless to say, older than civilization. It's even older than humanity, though humans are considerably more efficient than most draft animals in converting sunlight into work; an adult human is able to convert roughly 18 percent of the calories he consumes into work, while a big hayburner like a horse or ox is lucky to hit 10 percent—one of the reasons for the popularity of slavery throughout history. The remaining two were, by the seventeenth century, relatively mature technologies. More than 3,500 years ago, Egyptians were using waterwheels both for irrigation and milling, while at the other end of Asia, first-century Chinese engineers were building waterwheels linked to a peg and cord that operated an iron smelter's bellows; the following century, another Chinese waterwheel used a similar mechanism for raising and dropping a hammer for milling rice. The first-century historian known as Strabo the Geographer described a water-driven mill for grinding grain in the palace of Mithridates of Pontus that was built in 53 BCE. A century later, the Roman architect and engineer Vitruvius (the same one who inspired Leonardo's "Vitruvian Man") designed, though possibly never built, a water mill that used helical gearing to turn the rotation of the wheel into the vertical motion of the grinding stone, "the first great achievement in the design of continuously powered machinery."

Europeans had been putting water to work for hundreds of years before they started harnessing the wind, possibly during the tenth century, and certainly by the twelfth. This was probably due to the fact that, even more than waterwheels, the utility, and therefore the ubiquity, of windmills was a function of geography. They were, for example, common in northern Europe, because of the flat topography, and their comparative advantage in a climate where rivers freeze in the winter, but rare enough in the Mediterranean that Don Quixote could still be astonished by the appearance of one.

Windmills and waterwheels were, and are, used for everything from pumping water to sawing wood to operating bellows to smoothing (the term of art is *fulling*) wool. As we will see, it was a century before steam engines were used for a function that was not previously, and usually simultaneously, performed by wind or water mills. Their most important function, however, from antiquity forward, was milling grain; in the case of muscle power, the same grain used for feeding the draft animals themselves.

Whatever their productivity in milling grain, wind and water mills suffered from two fundamental liabilities, one obvious, the other less so. The first was the fact that water mills, especially, are site-specific; work could be performed by them only alongside rivers and streams, not necessarily where the work was needed. The second, however, proved to be the more significant: The costs of wind and water power were largely fixed, which profoundly reduced the incentives to improve them, once built, any more than someone who bought a car that came with a lifetime supply of gas would seek to drive economically. One can make a water mill more powerful, but one cannot, in any measurable way, reduce its operating expenses. The importance of this can scarcely be underestimated as a spur to the inventive explosion of the eighteenth century. So long as wind, water, and muscle drove a civilization's machines, that civilization was under little pressure to innovate. Once those machines were driven by the product of a hundred million years of another sort of pressure, innovation was inevitable. One is even tempted to say that it heated up.

COAL IS SUCH A critical ingredient for the Industrial Revolution that a significant number of historians have ascribed Britain's industrial preeminence almost entirely to its rich and relatively accessible deposits. Newcomen's engine, after all, ran on coal, and was used to mine it. One would scarcely expect to read a history of the steam engine, or the Industrial Revolution, without sooner or later encountering coal.

Encountering it in the same chapter that documents the rise of the experimental method is perhaps a little less obvious. But that proximity is neither sloppiness nor coincidence; the two are subtly, but inextricably, linked. The mechanism by which the steam engine was first developed, and then improved, was a function not only of a belief in progressive improvement, but of an acute awareness that incremental improvements could be measured by reducing cost. Demand for Newcomen's steam engine was bounded by the price of fuel per unit of work.

For a million years, the fuel of choice for humans was hydrocarbons, in the form of both wood and charcoal, but it did no work, in the mechanical sense. Instead, it was used exclusively to cook food and combat the cold, and, occasionally, to harden wood. Several hundred thousand years later, a group of South Asians, or possibly Middle Easterners, discovered that their charcoal fires also worked pretty well to turn metals into something easier to make into useful shapes, either by casting or bending. For both space heating and metalworking, wood, the original "renewable" fuel, was perfectly adequate; measured in British Thermal Units—

as above, the heat required to raise the temperature of one pound of water 1°F—a pound of dry wood produces about 7,000 BTUs by weight, charcoal about 25 percent more. Only as wood became scarce did it occur to anyone that its highest value was as a construction material rather than as a fuel. It takes some fourteen years to grow a crop of wood, and burning it for space heating or for smelting became a progressively worse bargain.

Europe's first true "wood crisis" occurred in the late twelfth century as a bit of collateral damage from a Christian crusade to destroy the continent's tree-rich sanctuaries of pagan worship and open up enough farmland to make possible the European population explosion of the following centuries. A lot more Europeans meant a lot more wooden carts, wooden houses, and wooden ships. It also meant a lot more wood for the charcoal to fuel iron smelters, since smelting one pound of iron required the charcoal produced by burning nearly eight cubic feet of wood. By 1230, England had cut down so many trees for construction and fuel that it was importing most of its timber from Scandinavia, and turned to what would then have been called an alternative energy source: coal.

Coal consists primarily of carbon, but it includes any number of other elements, including sulfur, hydrogen, and oxygen, that have been compressed between other rocks and otherwise changed by the action of bacteria and heat over millions of years. It originates as imperfectly decayed vegetable matter, imperfect because incomplete. When most of the plants that covered the earth three hundred million years ago, during the period not at all coincidentally known as the Carboniferous, died, the air that permitted them to grow to gargantuan sizes—trees nearly two hundred feet tall, for example—collected its payback in the form of corrosion. The oxygen-rich atmosphere converted most of the dead plant matter into carbon dioxide and water. Some, however, died in mud or water, where oxygen was unable to reach them. The result was the carbon-dense sponge known as peat. Combine peat with a few million years, a few thousand pounds of pressure, several hundred degrees of heat, and the ministrations of uncounted billions of bacteria, and it develops through stages, or ranks, of "coalification." The shorter the coalification process, the more the final product resembles its plant ancestors: softer and moister, with far more impurities by weight. Or so goes the consensus, "biogenic" view of coal's natural history; an alternative theory does exist, arguing that coal and other fossil fuels have a completely geological origin. The theory, arrived at independently in the 1970s by Soviet geologists and the Austrian-born astrophysicist Thomas Gold,* contends that the pressures and heat present in what Gold called the "deep hot biosphere" formed the hydrocarbon fuels currently being used to run the world's energy economy, and the presence of biological detritus in coal and other "fossil" fuels is a side effect of the bacteria that fed on them.

Whether as a result of geologic or biogenic forces, each piece of coal is unique, the result of both different plant origins and differing histories of pressure, heat, and fermentation. What they have in common is that they all share the same relationship between time and energy: Over thousands of millennia, hydrogen and hydroxyl compounds are boiled and pushed out, leaving successively purer and purer carbon. The younger the coal, the greater the percentage of impurities, and the lower the ranking. In fourteenth-century Britain, lower-ranked minerals like lignite and sub-bituminous coals were known as "sea coal," a term with an uncertain etymology but whose likeliest root is the fact that the handiest outcroppings were found along seams leading along the River Tyne to the North Sea.

Long before concerns about particulate pollution and global warming, coal had PR problems. Almost everyone in medieval England found the smell of the sea coal obnoxious, partly because of sulfuric impurities that put right-thinking Englishmen in mind of the devil, or at least of rotten eggs; by the early fourteenth century, it was producing so much noxious smoke in London that

* Gold died in 2004 after four decades at Cornell University and a lifetime of swimming outside the mainstream of scientific orthodoxy. In the 1950s, along with Fred Hoyle, Gold was the originator of the so-called steady state theory of the universe, which preceded and contradicted the generally accepted big bang theory.

King Edward I forbade burning it, with punishments ranging from fines to the smashing of coal-fired furnaces. The ban was largely ignored, as sea coal remained useful for space heating, though distasteful. Working iron, on the other hand, required a much hotter-burning fuel, and in this respect the softer coals were inferior to the much older, and harder, bituminous and anthracite. Unfortunately, along with burning hotter and cleaner—a pound of anthracite, with a carbon content of between 86 and 98 percent by weight, produces 15,000 BTUs, while a pound of lignite (which can be as little as one-quarter carbon) only about 4,000 to 8,000 BTUs—hard coal is found a lot deeper under the ground. Romans in Britain mined that sort of coal, which they called *gagate,* and we call jet, for jewelry, but interest in deep coal mining declined with their departure in the fifth century. It was not until the 1600s that English miners found their way down to the level of the water table and started needing a means to get at the coal below it.

ANY NARRATIVE HISTORY OF the steam engine must sooner or later make a detour underground. An Industrial Revolution without mining, and particularly coal mining, is as incomplete as rock and roll without drums. Actually, that understates the case; the degree to which cultures have solved the geometric puzzle of extracting useful ores from the complicated and refractory crust of the earth is a pretty fair proxy for civilization itself.

At more or less the same moment in history that groups of *H. sapiens* started digging a few inches into the dirt in order to plant seeds, they also began digging a little farther looking for flint that could be chipped into useful shapes. Even earlier, some forty thousand years ago, the caves at Lascaux were decorated with, among other things, pigment extracted from the iron ore known as hematite. Sometime thereafter, every community of humans discovered that clay plus fire equals pottery, and, by about 4000 BCE, that the earth contained semiprecious stones like turquoise and malachite, and easily worked metals like gold, and especially copper. The Roman mines along the Rio Tinto (named for the color its copper ore gave the water) in southern Spain not only provided precious metals but mechanized the process for the first time, using aqueducts to wash the debris out of the excavation and waterwheels to crush the ore left behind. And, of course, for four centuries Roman Britain was a source of silver, copper, and gold—and jet—for the imperial treasury.

The demand for Roman engineering was a function of the change from surface to deep mining, though the adjective is a relative one. The deepest mine in the world, at 2.4 miles down, is the Tau Tona gold mine outside of Johannesburg; the world's deepest coal mine is barely three-quarters of a mile to the bottom, which means that the most elaborately dug structures in mining history have scratched only the tiniest fraction of the mineral trove of the planet.*

Despite six millennia of improvement in mining technology, the "scratching" is actually more dangerous today than it was in the Neolithic period, and nearly as hazardous as it was for medieval pick-and-axe miners. Once the potential for surface mining, which is the complete removal of the ground cover, was exhausted, the only recourse for coal extraction was digging, usually into a hilltop. Whether the goal was hard coal or soft, the first step in such digging was mounting a large bore auger on a framework, rotating it, usually by either men or tethered mules walking in circles, and adding segments to it as it drilled deeper. The auger was followed by miners using tools, primarily picks, to carve coal from seams (some up to 100 feet thick) in a "room-and-pillar" method at the face, and transporting the coal by cart to the adit, or borehole entrance. With increasing depth, water-driven elevators or skip buckets were used to carry coal to the surface. In medieval England, the combination of technical difficulty with the ever-present risk of cave-in, flooding, and sharp tools wielded in close

* A back-of-the-envelope calculation, using 33×10^{12} cubic miles as the rough spherical volume of the planet, concludes that excavating a truly giant mine—2.4 miles down, a mile on a side—gets at no more than 7×10^{-9} of the earth's volume. Barely a scratch.

quarters meant that miners were treated like a relatively privileged class; unlike tenant farmers, they dug without obligation to the lords whose lands they worked, living "in a state within a state, subject, only, in the last resort, to the approval of the Crown."

As coal mines went deeper, they also became more dangerous, and not merely because of the engineering challenge of supporting tons of overburden; one of the volatile components of raw coal is the hydrocarbon CH_4, or methane, which is the main component of the flammable mixture known as "firedamp." Though it is slightly lighter than air, it can still pool in sealed areas of mines, causing a danger of asphyxiation and, far more significant in an age in which the only illumination came from fire in one form or another, explosion.* Savery's "Miner's Friend" was not, as it happens, sold exclusively as a water pump, but also as a means for ventilating such mines.

Anything that improved mining was attractive to the innovators of eighteenth-century England. Three-quarters of the patents for invention granted prior to the Savery engine were, one way or the other, mining innovations; 15 percent of the total were for drainage alone, as the shortage of surface coal became more and more acute and prices rose.

Price is the mechanism by which we allocate the things we value, from iPhones to coal, and even an imperfect system sooner or later incorporates the cost of manufacture into the selling price. In 1752, a study was made of a 240-foot-deep coal mine in northeast England in which a horse-driven pump lifted just over 67,000 gallons every twenty-four hours at a cost of twenty-four shillings, while Newcomen's engine pumped more than 250,000 gallons using twenty shillings' of coal—a demonstration not only of the value of the engine, but of a newfound enthusiasm for cost accounting. Newcomen's engine, by pumping water out of deeper mines at a lower cost, also lowered the effective price.

* An explosion is essentially a fast-burning fire with nowhere to go. Firedamp, however, can also burn slowly. *Very* slowly. The mine fire that started in Centralia, Pennsylvania, in May 1962 is, as of this writing, still burning.

The problem was that it didn't lower it enough. The coal-fired atmospheric engines of the type designed by Newcomen and Calley burned so much coal for the amount of water they pumped that the only cost-effective place for their use was at the coal mine itself. This did a lot more for heating British homes than running British factories; as late as the 1840s, the smoky fireplaces of British homes still consumed two-thirds of Britain's domestic coal output, and a shocking 40 percent of the world's. An eighteenth-century coal porter in London might carry loads of twice his own weight up rickety stairs and even ladders up to sixty times a day. But no one was using steam engines for much else, because the cost of transporting the coal to a steam engine more than a few hundred yards from the mine itself ate up any savings the engine offered.

For fifty years, lowering the cost of mining coal for heat had been enough to make the Newcomen engine a giant success. It was dominant in Britain, copied all over Europe, and even studied at universities—unsurprisingly, given the experimental methods that had created the engine in the first place. One of the universities interested in producing a superior version of the Newcomen engine was the University of Glasgow, the fourth oldest in the English-speaking world, and home not only to Joseph Black but to James Boswell, Adam Smith, and a dozen other leading lights of what came to be known as the Scottish Enlightenment.

And, of course, to James Watt.

CHAPTER FIVE

SCIENCE IN HIS HANDS

concerning the unpredictable consequences of sea air on iron telescopes; the power of the cube-square law; the Incorporation of Hammermen; the nature of insight; and the long-term effects of financial bubbles

THE FINEST ANCHORAGE IN the Caribbean is found on the southeastern coast of Jamaica, behind an eight-mile-long sandbar that protects the harbor from tropical storms. At the western tip of the sandbar, the original Spanish colonizers built a town they called Santiago, and which the island's English conquerors subsequently renamed Port Royal, retaining it as a base for privateering until it was destroyed by an earthquake in 1692. Kingston, on the mainland side of the harbor, was built as a refuge for survivors of the earthquake. It proved an attractive destination for refugees of another disaster, the Jacobite Rebellions of the early eighteenth century (the fruitless attempts to return the Stuart kings to the throne following the Glorious Revolution of 1688), which resulted in, among other things, the emigration of thousands of Scots to the island.* In 1747,

* Place names like Aberdeen and Culloden testify to the Scottish influence on Jamaican history.

one of them (a Scot, not a Jacobite), Alexander Macfarlane, a merchant, a judge, a mathematician, and yet another of those "gentlemen, free and unconfin'd" who could style themselves Fellows of the Royal Society, acquired several dozen state-of-the-art astronomical instruments from another Scot named Colin Campbell. Campbell was not merely a countryman, but a fellow alumnus of Glasgow University, so it was scarcely surprising that when Macfarlane died in 1755, his collection was bequeathed to their alma mater.

The ships that traveled from the Caribbean to Britain had a good deal more experience carrying sugar than they did telescopes and quadrants, whose iron components were not improved by several weeks exposure to salt air. Which is why, when the Macfarlane collection arrived in Glasgow in 1756, the university hired an artificer, just returned from London, "to clean them and to put them in the best order for preserving them from being spoilt."

James Watt was then twenty years old, and events had been preparing him for his new job almost since he was born, in Greenock, a borough just to the west of Glasgow on the River Clyde. Or even before. Scotland had formally joined the United Kingdom in 1707, but remained distinct from its southern neighbor in a number of relevant ways: poorer, but more literate, and far less inhibited by the presence of an established church that was turning Oxford and Cambridge into vocational schools for the clergy. The combination of relative poverty, and opportunity in British possessions around the world, explained the particularly Scottish enthusiasm for education: if the nation's most ambitious and smartest sons had to seek their fortunes elsewhere—and they did; during the eighteenth century, as many as six thousand trained Scottish doctors left the country in search of employment, and not just in Jamaica—the most valuable property they could take with them was between their ears.

As a result, even members of Scotland's artisan classes were better educated than was the case almost anywhere else in Europe—a bit of good fortune for them, but even more so for Britain's ability to maintain its head start on the development of steam power. A 1704 *Proposal for the Reformation of Schools and Universities* proposed a curriculum in mathematics that would seem daunting to a twenty-first-century honors student: "the first six, with the eleventh and twelfth Books of *Euclid,* the Elements of *Algebra,* [and] the Plain and Spherical *Trigonometry*" followed by "The Laws of *Motion, Mechanicks, Hydrostaticks, Opticks* . . . and *Experimental Philosophy.*" Watt, in particular, was taught an impressive amount by a cousin: "John Marr, *mathematician,*" as he appeared in Greenock's census.

Like Marr, Watt's grandfather had been a teacher of mathematics, navigation, and astronomy; his father was a carpenter specializing in shipbuilding who supplemented his income by surveying the land around Greenock, but both were famed for their skill in the repair of delicate instruments. And so, therefore, was James, though whether his combination of mathematical and mechanical aptitude was genetic or the result of early training is as unknowable as it is irrelevant, since all memoirs of Watt's childhood suffer from the sort of retrospective adulation that nations habitually bestow on their heroes' early years. Watt certainly seems to have been a bright and precocious boy, but his childhood history is decorated by a truckload of conveniently postdated reminiscences (see Cherry Tree, George Washington's). In Watt's case, the best one is the story of his aunt's recollection of young James's obsession with the way a teakettle lid was forced upward by steam—suspicious on the face of it, since, as we have seen, the expansive force of steam was not precisely central to the operation of early steam engines. In any event, he certainly benefited from being given the full run of his father's workshop, with its hammers, chisels, adzes, block and tackles, and so on.

When Watt's mother died in 1753, the seventeen-year-old was sent to Glasgow to learn the trade of a "mathematical instrument maker," and though he could find no teacher, he did eventually encounter Robert Dick, a doctor and the professor of natural philosophy at the University of Glasgow. Dick was unable to provide training, but he did advise Watt to seek a teacher in London, for which he supplied a letter of introduction. Taking both the advice and the letter, Watt left Scotland on June 7, 1755, arriving in London twelve days later.

The city, then home to more than 600,000 residents, was already the largest outside of Asia, and easily the dirtiest. Though London owes much of its finest architecture to the fire of 1666, which cleared the way for the buildings of Christopher Wren and Robert Hooke, the overwhelming bulk of the city's buildings were constructed to somewhat lower standards than St. Paul's. Moreover, it was still, as of the date of Watt's arrival, using the Thames for both sewage discharge and drinking water, which partly explains why so much of poor London slaked its thirst with gin, a distilled spirit made from fermenting grain that was so bad it couldn't even be used to make beer. The Hogarthian enthusiasm for the cheap liquor was such that Henry Fielding—novelist, dogooder, and pioneer of London's first police force—wrote, "it is the principal sustenance (if it may be so called) of more than a hundred thousand people in this metropolis. Many of these Wretches there are, who swallow Pints of this Poison within the Twenty Four Hours: the Dreadfull Effects of which I have the Misfortune every Day to see, and to smell too." With its large and unwashed populace, its untreated sewage, and the miasma caused by burning nearly two-thirds of the world's output of decidedly dirty coal, the city literally stank.

The smells were part of the cost of supporting the world's most robust commercial and manufacturing economies, but while the former was dominated by newly created speculative ventures, funds, and trading syndicates, the latter had a more medieval flavor. In London, as in most cities of Europe, the making of things had long been the prerogative of guilds, those ancient federations of autonomous workshops whose grip on activities such as weaving cloth, making jewelry, and working metals imposed very substantial costs on the city's economy.

Some of those costs were borne by the guilds' prospective membership in the form of free labor and apprentice fees, paid in return for both training and a de facto license to practice the skills acquired. The training, of course, is what James Watt had traveled to London to acquire, from the city's Worshipful Company of Clockmakers. That particular guild was not a true medieval organization; it had been founded "only" in 1631, just in time to define its exclusive franchise as embracing not only clocks but all forms of mathematical instruments. Partly as a result, they were considerably more welcoming of innovation than the more ancient organizations; when Watt arrived in London, their most illustrious member, John Harrison, was not only improving on his prizewinning marine chronometer, which he had invented as a solution to the problem of calculating longitude at sea, but also had previously created new versions of both clock escapements and pendulums. Unfortunately for Watt, Harrison's guild was just as jealous of their territorial prerogatives as any thirteenth-century goldsmith; their bylaws prohibited any member from employing—and, especially, training—any "foreigners, alien or English" unless they were bound to the member as apprentices.

As a result, the first thing Watt learned in London was that he did not qualify for a "normal" apprenticeship. He was too old, for one thing. And even had he been closer to the usual age of apprentices, he had no interest in spending seven years as one. On the other hand, his willingness to leave London once trained was a huge advantage, since the guild rules were explicitly designed to eliminate unauthorized competitors only within the city. He was also, by training and aptitude, already far more useful to a master clockmaker than a fourteen-year-old still picking hay out of his ears. The combination was evidently appealing enough that John Morgan, a member of the Company in good standing, agreed to take Watt on as a trainee in return for a year of free labor plus twenty guineas. By all accounts, he got a bargain: Since his "apprentice" had neither an interest in frivolity, nor the funds to indulge one, he did nothing but work. Watt was attempting to crowd seven years of training into one, and he succeeded. Most of his training was in fine brasswork, building sectors, dividers, and compasses; even a Hadley quadrant with a telescope and three mirrors. He boasted to his father that he had mastered an extremely precise "French joint"—a hinge in which one channel folds into another like a fine bound book. By the time he returned to Glasgow in 1756, he was certified "to work as well as most journeymen" and was qualified to build and repair the machines representing the eighteenth century's most advanced technology.

Glasgow was then barely a town by London standards, home to around fifteen thousand people, but it was a "large, stately, and well-built city . . . one of the cleanliest, most beautiful, and best-built cities in Great Britain" in the words of Daniel Defoe,* who visited in 1724 to report on Scotland's integration with England. It was also, in Defoe's words, "a city of business [with] the face of foreign as well as domestick trade" and a textile manufacturing center specializing in "stuff cross-striped with yellow, red, and other mixtures" (i.e. plaid), which meant that it was also home to its own guilds, just as jealous of their prerogatives as their London counterparts. In the case of Watt's newfound skills, the barrier to entry was manned by the rather fearsome-sounding "Incorporation of Hammermen," who, in the time-honored practice of every guild, weren't enthusiastic about recognizing a competitor who had failed to go through an approved apprenticeship. So when his former patron, Professor Dick, in need of someone to repair the sea-damaged Macfarlane collection, offered a payment of £5, and more important, permitted him to set up shop as "Mathematical Instrument Maker to the University," it was truly a godsend.

It is almost irresistibly tempting to see Watt's life as the embodiment of the entire Industrial Revolution. An improbable number of events in his life exemplify the great themes of British technological ascendancy. One, of course, was his early experience with the reactionary nature of a guild economy, whose raison d'être was the medieval belief that the acquisition of knowledge was a zero-sum game; put another way, the belief that expertise lost value whenever it was shared. Another, as we shall see, was

* It wasn't Defoe's first comment on the new world being created in Britain. In his 1697 *Essay on Projects,* he named his era "the Projecting Age," by which he meant the "projectors" who sought to build commercial empires supported by patents and monopolies (and, to be fair, "projects" like overhyped investments, about which more below).

his future as the world's most prominent and articulate defender of the innovator's property rights. But the most seductive of all was Watt's simultaneous residence in the worlds of pure and applied science—of physics and engineering. The word "residence" is not used figuratively: The workshop that the university offered its new Mathematical Instrument Maker was in the university's courtyard, on Glasgow's High Street, a bare stone's throw from the Department of Natural Philosophy.*

He almost immediately started collecting admirers. One of his first friends among the university's "natural philosophers" was the mathematician and physicist John Robison, who was therefore in a privileged position to observe Watt in the years before his great achievements. Nearly forty years later he would recall that "every thing became Science in [Watt's] hands . . . he learned the German language in order to peruse Leopold's *Theatricum Mechanicum* [an encyclopedia of mechanical engineering] . . . every new thing that came into his hands became a subject of serious and systematical study, and terminated in some branch of Science." He continued:

> Allow me to give an instance. A Mason Lodge in Glasgow wanted an Organ [and] tho' we all knew that he did not know one musical note from another, he was asked if he could build this Organ. . . . He then began to study the philosophical theory of Music. Fortunately, no book was at hand but the most refined of all, and the only one that can be said to contain any theory at all, Smith's *Harmonics*. Before Mr. Watt had half-finished this Organ, he and I were completely masters of that most refined and beautiful Theory of the Beats of imperfect Consonances. He found that by these Beats it would be possible for him, totally ignorant of Music, to tune this Organ according to any System of temperament, and he did so, to the delight and astonish-

* And only yards away from the Department of Moral Philosophy, where Adam Smith, whom we will meet in chapter 11, had held a professorship since 1751.

> ment of our best performers. . . . And in playing with this he made an Observation which, had it then been known, would have terminated a dispute between the first Mathematicians of Europe, Euler and d'Alembert, and which completely establishes the theory of Daniel Bernoulli about the mechanism of the vibration of Musical Chords. . . .

Watt may have been comfortable in the rarefied company of mathematicians like Bernoulli and Leonhard Euler; the business alluded to by Robison is the discovery that any of the overtones of an organ pipe produce frequencies that are exact multiples of the pipe's base pitch. However, like Newcomen (but unlike Boyle, or even Savery), he was as preoccupied by his desire to earn a living as by his passion for discovery. Like an ever-growing percentage of his countrymen in the newly United Kingdom, Watt had acquired the tools necessary for scientific invention—the hands of a master craftsman, and a brain schooled in mathematical reasoning—without the independent income that could put those tools to work exclusively for the betterment of mankind. As a result, in 1759, Watt became half of a partnership with John Craig manufacturing optical instruments. In 1763, he became shareholder in the Delftfield Pottery Company. And every year, he spent a portion of the spring and summer working as a surveyor for the roads and canals just starting to crisscross Britain.

It was upon his return from a surveying trip, in the winter of 1763, that Watt was asked to repair a model of a Newcomen engine in the possession of the university by John Anderson,* who had become Glasgow's professor of natural philosophy with the death of Robert Dick in 1757. "Repair" is something of a mis-

* Anderson, whose nickname among the university's students was "Jolly Jack Phosphorus," is a fascinating character in his own right: A professor of Hebrew and Semitic languages as well as natural philosophy, he is best remembered as an early advocate of higher education for artisans and craftsmen, for whom he held classes throughout his forty years at Glasgow. So dedicated was he to this underutilized national resource that his estate was used to found Anderson College, now part of the University of Strathclyde.

nomer; the model was not broken, but unlike a full-sized engine, it stopped working after only two or three strokes. Anderson had been importuning the university's new instrument maker for at least four years before Watt "set about repairing it as a mere mechanician." Shortly thereafter, he realized that the problem was intrinsic to the size of the model, since "the toy cylinder exposed a greater surface to condense the steam in proportion to its content." Watt had intuited the presence of a cube-square problem.

The so-called cube-square law is a recognition of the fact that the surface of any solid object increases in size far more slowly than its volume. Thus, a cube with four-inch sides has a surface area of ninety-six square inches and a volume of sixty-four cubic inches, while an eight-inch cube has a surface of 384 square inches, but a volume of 512 cubic inches. Doubling the cube's edge increases its surface area fourfold, but its volume eight times.

The cube-square law is yet another bequest from the Scientific Revolution to the Industrial; for a change, one with a clear provenance. The phenomenon was first documented in the final book of Galileo Galilei, the 1638 *Dialogues Concerning the Two New Sciences,* in which Galileo's alter ego, the imaginary "Salviato," demonstrates it to the Aristotelian loyalist "Sagredo" and the dim-minded "Simplicio" (Galileo's choice of names was as heavy-handed as Dickens's). The cube-square law has huge implications for construction, for engineering, and even for biomechanics; it is the reason, for example, that an elephant's legs are so much larger in cross-section than a dog's. More relevantly for the history of steam power, it reveals the most obvious weakness of scale models, which is that a structure's performance can degrade substantially when it is blown up to twenty times its original size. Designs that work when small—a bridge made of toothpicks, for example—can easily fail as the weight to be borne increases disproportionately faster than the strength of the "timbers" bearing it.

But the problem also operates in reverse. The cube-square law can just as easily cause a design to fail when it is miniaturized. This was Watt's initial insight about the model Newcomen engine. Because the scale model, still in the Hunterian Museum at the university, was using far more steam than could be accounted for by any science or experience Watt (or anyone else) had, his first assumption was that the problem was one of the scale itself, specifically the fact that in a small engine the interior surface was far larger in proportion to the volume; if the heat loss was proportional to surface, then the difference could perhaps be explained.

Explaining it took two years.

Watt's experiments from 1763 to 1765 were an object lesson in the primacy of measurement over intuition, since recognizing the *existence* of heat loss matters a good deal less than knowing its *magnitude;* suspecting the nature of the problem wasn't the same as understanding it. Watt needed to calculate exactly how much heat was being lost in the Newcomen design, and that meant converting general theories about steam into precise measurements, which were, to be kind, thin on the ground at the time, even for such elementary benchmarks as the boiling point of water.

Obviously, the story of steam demands constant reference to that benchmark, which even a bright ten-year-old knows is precisely 212°F, or 100°C, at normal atmospheric pressure. However, as with many such bits of common knowledge, it turns out to be a bit more complicated. Boiling occurs when a liquid's vapor pressure reaches atmospheric pressure, but while vapor pressure is proportional to heat, it isn't the same throughout a volume of liquid. Boiling temperatures change depending on the material containing the liquid, since water adheres better to metal than to glass and can therefore boil at a somewhat lower temperature in a metal vessel. The temperature can increase or decrease with the shape of the container, the presence of dissolved air, the location of the heat source, and, of course, the amount of air pressure. Thus, the "normal" boiling temperature of water—100°C—can climb as high as 200°C, as an obsessively competitive scientist named Georg Krebs demonstrated in 1869. Most textbooks plot a "boiling curve" with the boundary between liquid and gas a moving target depending on at least four different variables.

Even in Watt's time, the clear line between liquid and vaporized water was pretty fuzzy. A thermometer dating from the 1750s is marked with two different "boyling" temperatures; at 204, water "begins to boyle," and then at 212, "boyles vehemently, a distinction that dates back to Isaac Newton. The measurement problem was acute enough that in 1776 the Royal Society appointed a committee, headed by Henry Cavendish (better known as the discoverer of hydrogen) in order to establish the "fixed points" of thermometers.

Watt began his researches on the Newcomen engine fourteen years before the Cavendish committee delivered its conclusions in 1777 and was, in consequence, working with a clunkier set of measurements. He wasn't, however, completely in the dark. Toward the end of his life, Watt himself provided an inventory of the basic knowledge already in circulation before his first great innovation. One small example of it was the twenty-year-old discovery, by the physician William Cullen (Joseph Black's teacher, and yet another member of the remarkable faculty of the University of Glasgow), that water boiled at a lower temperature in a vacuum, thus releasing steam that would degrade the cylinder's vacuum. This turned out to be critical, because the fact that the Newcomen engine operated in a vacuum meant that cooling the steam to the point of condensation required cooling it to temperatures even lower than 100°C/212°F. In order to calculate how much lower, Watt needed to develop an exact scale showing how changes in pressure map to boiling temperatures. Most important, he needed an accurate way of measuring the volume of steam produced by vaporizing a given volume of water, and the water condensed from a measured amount of steam. Watt was a demon for measurement, and he spent months computing the volume of steam as compared to water, the quantity of steam used by a single stroke of a Newcomen engine, the quantity of water needed to condense it, and so on. As a case in point, though Samuel Moreland had estimated that boiling a given volume of water would produce steam that would fill a space 2,000 times greater, J. T. Desaguliers calculated the number as 14,000, and Watt needed to find out for himself. In one of his notebooks, he describes an experiment in which he boiled an ounce of water in a "Florence Flask," forced the air and water out, and compared before and after weights, concluding that the accurate relationship between liquid and solid volumes was 1,849 times. Once deriving the critical relationship between the phases of water, as Watt later recalled,

> I mentioned it to my friend Dr. Black, who then explained to me his doctrine of latent heat. . . . I thus stumbled upon one of the material facts by which that beautiful theory is supported. . . . Although Dr. Black's theory of latent heat did not *suggest* my improvements on the steam-engine . . . the correct modes of reasoning, and of making experiments of which he set me the example, certainly conduced very much to facilitate the progress of my inventions.

Nothing is more common in the history of science than independent discovery of the same phenomenon, unless it is a fight over priority. To this day, historians debate how much prior awareness of the theory of latent heat was in Watt's possession, but they miss Black's real contribution, which anyone can see by examining the columns of neat script that attest to Watt's careful recording of experimental results. Watt didn't discover the existence of latent heat from Black, at least not directly; but he rediscovered it entirely through exposure to the diligent experimental habits of professors such as Black, John Robison, and Robert Dick.

In the end, it was the habits of recording and comparing results, time after time, that proved truly indispensable for Watt's "rediscovery" of Joseph Black's conjecture of latent heat, one that puzzled not only Watt, but generations of physics students ever since. Boil a quart of water, turning it into steam; it takes up a bit more than 1,800 times the space it did when liquid. But an atmospheric steam engine doesn't want steam, it wants a vacuum, so it has to condense that steam back into water. Newcomen did so by injecting a stream of water into the sealed cylinder of his engine, but he never measured the amount needed. Watt, diligent experimentalist that he was, did: It took up to six quarts of water at room temperature to condense the steam. A year into the process, Watt had not only rediscovered Black's theory, he was finally able to quantify it. The exhaustive process of experimenting, measuring, and experimenting again had allowed him to calculate how much steam was necessary for each piston stroke, and how much the Newcomen engine was actually generating. He now had quantities he could measure.

The measurements showed him where the problem was. The engine depended on steam's filling the cylinder before it was ready to produce a vacuum. But every time fresh steam was admitted into the now cooled cylinder, it didn't expand; it just continued to condense, turning back into water until the cylinder heated up to the temperature of the steam itself. Heating the cylinder walls wasted up to three-quarters of the steam, or even more: In one test made in 1765, Watt found that an old-style engine was boiling more than three times as much water to heat the cylinder as it was using to create a vacuum.

The problem was exacerbated by the fact of the vacuum itself, which effectively lowered the vaporization temperature of the water, in the same way that water boils at a much lower temperature at high altitudes: lower pressure, lower boiling temperature. The water, which needed to be heated to 100°C/212°F to boil at normal pressure, needed only half that to boil in a vacuum. And if water turns to steam at relatively lukewarm temperatures, then condensing it requires either a modest amount of very cold water (obviously impractical without refrigeration) or a huge amount at room temperature, which degraded the engine's efficiency even further.

The Newcomen engine was caught between fundamentally incompatible goals: The engine should use as little water as possible to condense the steam (in order to avoid cooling the cylinder), but as much water as possible to make sure that condensation occurred rather than more vaporization. Put another way: The cylinder needed to be kept at a constant 212°F/100°C (to avoid condensation that didn't create a vacuum) *and* it needed to be kept at a constant 100°F/45°C (to avoid vaporization). Watt, in Usher's terms, had perceived an "unsatisfactory pattern."

Still, after two years of measurement, analysis, and experiment, the unsatisfactory pattern was all he had. Frustrating though that was, he kept at it, to satisfy not merely his curiosity, but also his wallet; in his own words, his mind "ran on making engines *cheap* as well as *good.*" From the beginning, Watt recognized the problem in terms of wasted fuel, which meant wasted money, and therefore an opportunity. An idea that could significantly reduce that waste was clearly going to make someone rich. That someone didn't need to be a skilled artisan. He didn't have to live in a culture that had only recently articulated a property right in ideas and drafted legislation protecting that right. Scientists and philosophers, as we have seen, had been paving the way for centuries before Watt, or even Newcomen. Eighteenth-century Britain wasn't any more hospitable to their brilliant innovations than anywhere else; but it was a *lot* more hospitable to innovators who couldn't afford to invest years of their lives with no hope of material gain. Watt was brilliant, unusually so. But he was also emblematic of hundreds, soon to be thousands, of men like himself, each of them searching for a "eureka" moment.

ALFRED NORTH WHITEHEAD FAMOUSLY wrote that the most important invention of the Industrial Revolution was invention itself. A number of others compete for second place, but the insight that came to James Watt in the spring of 1765 has a lot of support. By then, he had tried dozens of different ways to find a cylinder that would both heat up and cool down rapidly, even trying different materials for the cylinder itself, experimenting with brass, cast iron, and even wood "soaked in linseed oil, and baked to dryness," each trial repeated half a dozen times. Nothing had worked, until he had his epiphany,* one he later described as the realization that since

* For more about the nature of that flash of insight, see chapter 6.

> steam was an elastic body it would rush into a vacuum, and if a communication were made between the cylinder and an exhausted vessel it would rush into it, and might be there condensed without cooling the cylinder. I then saw that I must get rid of the condensed steam and injection-water if I used a jet as in Newcomen's engine. Two ways of doing this occurred to me. First, the water might be run off by a descending pipe, if an outlet could be got at the depth of thirty-five or thirty-six feet, and any air might be extracted by a small pump. The second was to make the pump large enough to extract both water and air. . . .

What he had envisioned was simple enough: a second chamber, connected to the cylinder by a pipe, through which the steam would flow. When it arrived in the new chamber, already surrounded by cool water, the steam would condense, a vacuum would be formed, and atmospheric pressure would pull the piston down—but the cylinder in which the piston traveled could stay hot as a new jet of steam entered it. One chamber would stay cool, the other hot, each time the engine cycled.

The separate condenser would prove not only central to the development of steam power and the entire Industrial Revolution that ran on it, but also an utterly necessary step on the way to the very different sort of engine that powered *Rocket*. It is also, happily, a rich test case of the mutually reinforcing relationship between abstract theorizing and rule-of-thumb engineering. Literally rule-of-thumb: As with all mechanical inventions, the insight that inspired the separate condenser could be *visualized*, but it wasn't worth much of anything until it could also be *handled*—note the linguistic clue. The human eye can see things that don't yet exist, but making them requires the human hand, and it was now time for Watt to return to his university workshop and let his skilled hands turn his insight into a model.

That year training on brass compasses and quadrants in London now proved its worth. Within weeks, he had handcrafted all the components for an engine: two cylinders, one piped to a boiler

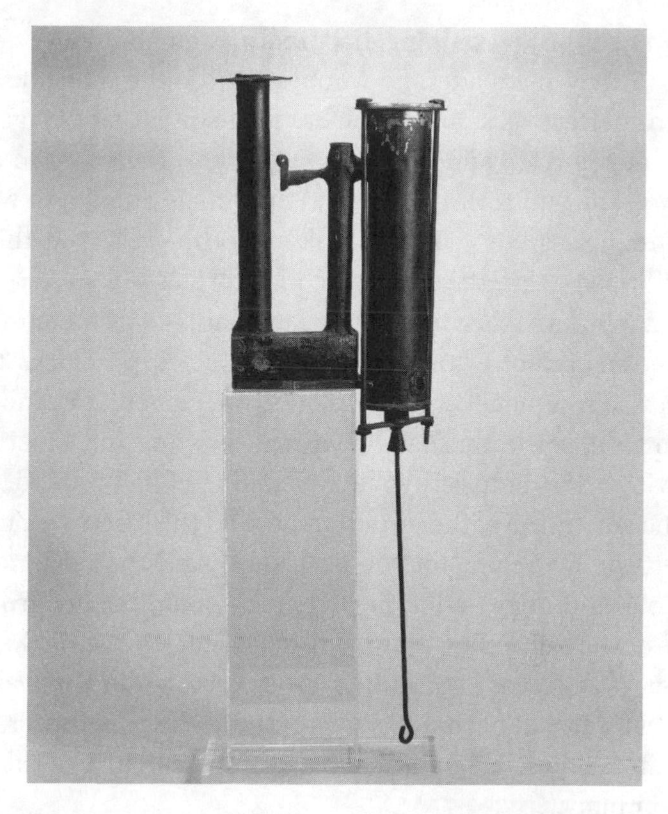

Fig. 3: The "stovepipe" on the left is Watt's separate condenser. Four years after his brainstorm on Glasgow Green, this was the result: a working cylinder that didn't need to be cooled and reheated for each stroke, thus doubling the utility of Newcomen's design. *Science Museum / Science & Society Picture Library*

and containing a piston with a valve on the bottom to vent excess water, the other a ten-inch-long brass syringe with a diameter of 1¾″ containing two ten-inch tin "straws" each about ⅙″ in diameter, and a hand-operated air pump with a ¾″ diameter.

The two cylinders were connected by a horizontal pipe, and the syringe immersed in a cistern of cold water. Watt lit his boiler and let the steam flow into the piston cylinder, closed the steam cock, and pumped out the air in the syringe, thus pulling in the steam, which immediately condensed around the cold tin straws. The piston in the cylinder immediately lifted a weight of 18 pounds; a cylinder holding barely a pint of water was raising a weight equivalent to more than two gallons. Watt, a perfectionist by temperament, education, and training, had finally (though briefly) satisfied himself. Thirty years later, he would describe the model as being "nearly as perfect . . . as any which have been made since that time." He was not normally an especially confident man—perfectionists rarely are—but in April 1765 he was optimistic enough to write to his friend James Lind, "I can think of nothing else but this Machine. I hope to have the decisive tryal before I see you. . . ."

It is not known when he actually saw Lind, but by the summer of 1765, on the back of £1,000 borrowed from Joseph Black, "the invention was complete . . . a large model, with an outer cylinder and wooden case, was immediately constructed, and the experiments made with it served to verify the expectations I had formed, and to place the advantage of the invention beyond the reach of doubt."

The time had come for the next step. And the next step was going to cost money—a lot more than he could borrow from colleagues at the university. Watt needed capital. Investment capital, however, wasn't easy to find in 1765 Britain; and it was a *lot* harder than it had been fifty years earlier. The reason was one of the greatest financial bubbles in history, the collapse of the South Seas Company.

THE SOUTH SEAS COMPANY had been incorporated in 1711, with a charter that granted what was potentially a far more lucrative monopoly than anything Edward Coke had contemplated a century earlier. In return for buying £10 million of government debt, the Company was given exclusive trading rights throughout Central and South America, whose bounty included wool, rum, sugar, and, most profitably, slaves. Promoted like an eighteenth-century Enron, the South Seas Company offered not only the promise of unimaginable wealth, but stock that could be purchased by virtually anyone. This was both a novel and an appealing idea in a time when the world's largest corporation, the British East India Company, had fewer than five hundred investors. Since, however, the Company's only real asset was the British government's promise of access to ports that were entirely controlled by the Spaniards, making money from trading proved difficult. The Company was, even so, brilliant at promoting its own prosperity, placing newspaper stories, hosting parties, and maintaining luxurious offices in the most expensive buildings in London. In January 1720, stock was issued at a par price of £100 a share; by August, at the peak of the bubble, when the average British artisan was earning less than £100 a year, a single share of the Company traded for £1,000. And even worse, it inspired other businesses to issue stock on what might be called speculative ventures, including a company capitalized at £1 million in order to produce a perpetual motion machine. One of the more candid styled itself "A Company for carrying on an undertaking of great advantage, but no one to know what it is."

The result, once the bubble burst and the dust cleared, was that Parliament essentially barred the issuing of stock for any business purpose, which limited the potential pool of investors to what would today be called venture capitalists. In the case of Watt's invention, this meant partnering with an entrepreneur with both ready cash and a liking for technology.

John Roebuck was then a forty-seven-year-old serial entrepreneur who had started half a dozen different businesses, each of them intending to exploit a technological innovation, including the first industrial refinery that manufactured sulfuric acid* by combining sulfur dioxide with oxygen and the resulting compound with water, all in a lead-lined chamber. The acid refinery prompted his first patent application, but by no means his last. By the time he met Watt in 1765, he was also master of one of the world's most innovative forges, the legendary Carron Ironworks, and holder of patent number 780—the number of patents granted

* An entire book—be undismayed, not this one—could be written on the history of sulfuric acid as a symbol for the evolution of modern civilization. Under the name "oil of vitriol," it was the most important weapon in the arsenal of medieval alchemists—the original philosopher's stone—and remains critical not only for producing fertilizer and bleaching textiles, but as a precursor chemical for sodium carbonate, which is essential for the manufacture of both paper and glass. Even now, a number of international economists use its production as a proxy for a nation's level of industrial development.

annually was still, a century after the Statute on Monopolies, measured in dozens—for a new process for making bar iron.

A correspondence between the ambitious young instrument maker and the nearly twenty years older businessman began in the summer of 1765, prompted by Joseph Black, who counted both as friends. In September, Watt wrote to Roebuck inviting an investment in his discovery that producing steam within a vacuum was dramatically more efficient than producing it in air, his excitement such that "I am going on with the Modell of the Machine as fast as possible and hope to have it finished in another week."

Rereading the letters, it is impossible to miss the tension present from the beginning. Roebuck fancied himself at least as gifted a scientist as Watt, and insisted on an extreme form of due diligence, demanding to see Watt's drawings, notes, and models. He demanded that Watt try to create a vacuum without a jet of water condensing the steam, and even urged him to discard the separate condenser, which was, after all, the point of the entire exercise. Watt, for his part, was generally courteous, but convinced of both his own talent and of the power of the separate condenser. For months, the two engaged in the sort of epistolary courtship that puts one in mind of the way that porcupines mate. Only when Roebuck, who was nothing if not intelligent enough to recognize Watt's gifts, satisfied himself that the separate condenser promised everything Watt believed, did he agree to a partnership. The terms of the agreement obligated Roebuck to absorb all future expenses related to building a machine that would, in Watt's words, produce the same amount of work for half the amount of fuel. He further agreed to pay off Watt's debt to Black in return for two-thirds of future profits.

Watt's frustrations were just beginning. Vacuum is notoriously unstable, but it needed to be kept intact in order for the engine to do its work. Newcomen's vacuum seal had been nothing more than a leather collar with a layer of water on top, but Watt had to avoid using it on his own engine, since the water would, of necessity, cool the hot cylinder and so eliminate most of the advantage of the separate condenser. But everything else he tried was either too porous—steam escaped *and* air entered—or created too much friction in the cylinder, costing a huge amount of energy. As a result, he tried dozens of combinations of materials for both piston and cylinder: wood, tin, copper, and cast iron, in square and round shapes, sealed with leather, cloth, cork, oakum, asbestos, and numerous alloys of lead, and lubricated with mercury, graphite, tallow, manure, and vegetable oil. "Cotton was proposed by my friend Chaillet; I thought of trying it but was deterred first by its price, secondly, by the very thing you have found: that it could not be easily made to cohere without glue or weaving the substances. I have hopes of pasteboard . . . mixed with dung; I propose to separate the gall and sand from the dung by washing. I have found by experiment that for making joints steam tight, there is nothing equal to it as it is of no consequence whether the joint be naturally round or not. . . ."

While pasteboard finally worked well enough, it did nothing to solve the central mechanical problem, which was getting the piston to fit into the cylinder as tightly as possible with as little friction as possible—as usual, two objectives fundamentally in conflict with each other. Over the first two years of experimentation, Watt built, again by hand, three models, each with a different cylinder: the original, with a cylinder of $1\frac{3}{4}''$ brass, but no steam jacket; a $1\frac{2}{5}''$ cylinder with steam jacket; and one five or six inches long with a steam jacket made of wood. The tin straws, which worked as a surface condenser, were discarded because of difficulties with consistency and replaced by jet condensers similar to those used by Newcomen.

Not all, or even most, of the revisions—in a nod to Usher's stages of invention, perhaps better to call them "critical revisions"—were the inspirations of a solitary inventor. A friend, Dr. William Small,* advised by letter, "Dear Jim . . . Let me suggest a

* Small would have been a key asset in any game of eighteenth-century "Six Degrees of Kevin Bacon" as a correspondent of Watt, a friend of Benjamin Franklin, and, before his return to Scotland from North America, Thomas Jefferson's onetime professor at the College of William & Mary.

method of making your wheel and valves tight: Let the valve frame be made easy for the groove and about half thick; put a ply of pasteboard below the frame . . . and place it in the groove in its proper place, then lay a ring of pasteboard all around each side of the groove and over each valve frame, taking care no pasteboard projects over the frames or grooves. . . ."

By 1768, Watt, three years into his deal with Roebuck, acknowledged that "what I knew about the steam engine [in 1765] was but a trifle to what I know now." His frustrations were growing pronounced. Any slight defect in any component was enough to compromise the design, and therefore the designer's temper. Newcomen's engine had only to be better than a horse-driven pump; Watt's had to be better than Newcomen's, and that meant cheaper. The unforgiving arithmetic of coal obliged him to produce not merely an elegant design, but one that consumed less fuel, and virtually anything less than perfection in the boring of the cylinder or the strength of the solder consumed more.

Watt's perfectionist habits, which had given rigor to his early experiments and made the original separate condensing model work so encouragingly, were no longer much of an advantage. Because while Watt could build a small model to the most exacting specifications, a larger, and therefore practical, version needed a design that could be executed by others; "my principal hindrance in erecting engines," he wrote to Roebuck in 1765, "is always the smith-work." Supposedly, the smiths at Roebuck's Carron foundry were the best in England, but even their skills were not up to making a cylinder to tolerances that resulted in one that was (a) perfectly round (so that the piston would fill it) and (b) airtight.

Watt's only "relief amidst [his] vexations" was, perversely enough, the need to make a living. Though Roebuck was paying the expenses while Watt was attempting to produce a working engine, and had even set the inventor up in a workshop at Kinneil, near the town of Borrowstounness (more popularly, Bo'ness) in central Scotland, he was not paying Watt a salary. To support his family—in 1764, Watt had married his cousin, Margaret Miller, who would give him five children before her death in 1772—the inventor adopted his father's trade, surveying the canals of northern England, which, he wrote, "have given me health and spirits beyond what I commonly enjoy at this dreary season. . . . Hire yourself to somebody for a ploughman; it will cure ennui."

At Kinneil, however, the pressure was unrelenting. By the middle of 1768, Watt had built an eighteen-inch cylinder out of tin, but the same malleability that made it an excellent material for sealing in the vacuum also made it something less than robust. Roebuck didn't care. He badly wanted some indication that his investment would be redeemed sometime soon, and he insisted that Watt apply for a patent. And so, in January 1769, Watt, somewhat reluctantly, traveled to London, where, despite the still imperfect design of his engine, he had been granted patent number 913 for "a method of lessening the consumption of steam and fuel in fire-engines." His first meeting after collecting the document from the Great Seal Patent Office was with neither Roebuck—the man who had financed the patent—nor Joseph Black, the friend who had inspired it, but with a Birmingham manufacturer named Matthew Boulton.

BOULTON WAS THEN THIRTY-NINE years old, eight years older than Watt, born into a family that made small metal goods: buckles, buttons, graters, household tools—"toys," in the vernacular of the day. When he was still in his teens, he entered the family business, most of whose functions were, typically for the time, jobbed out to others: raw materials were bought from one firm, sales handled by another, transportation by a third. Sometimes the other firms were dependable, sometimes not. But they were always costly, which seemed to Boulton an opportunity. By the time he was twenty-five, he had not only enlarged the business but was in the process of changing it irrevocably. Determined to integrate all possible aspects of manufacturing, Boulton moved the metal stamping operations from one water mill to another, starting construction in 1762 on what would become the world's largest and most famous factory with the relatively modest outlay of £9,000.* Eventually settling two miles from the center of the city of Birmingham, the Soho Manufactory would grow to include workshops, showrooms, stores, offices, worker dormitories, and design studios. It also incorporated a decidedly progressive bent in workforce relations: Boulton used no child labor, and he even offered his laborers, in return for one-sixtieth of their wages, social insurance that paid benefits in the event of illness or injury.

By the time he was thirty, he was already acknowledged as not only a visionary businessman, but also a hugely successful one. Soho's output of jewelry, silverware, and gilt decorative products, as well as the traditional iron and tin "toys," made Boulton, in the words of Josiah Wedgwood (himself a rather remarkable story in the history of ceramics), "the Most compleat Manufacturer of Metals in England." And he was more. James Watt is very likely the best known of all the inventors associated with the introduction of steam power. Partly this is because his life is such a useful bit of shorthand for the entire world of invention that fueled the perpetual innovation machine we call the Industrial Revolution. But the unique elements that made Britain so hospitable to inventions produced by her artisan class, including the legal and cultural incentives articulated in Coke's Statute and Locke's *Treatises,* were only half of the transaction. Increasing the supply of inventors by permitting them to sell their ideas was useless without a market in which those ideas could be sold. And since ideas don't sell themselves any better than anything does, someone needed to sell them. If James Watt was *primus inter pares* on the supply side of the steam economy, Matthew Boulton was unquestionably the man best equipped to introduce him to those willing to pay for his supply of ideas.

Watt had already visited the Soho Manufactory once before, in 1767. It is not known whether Watt, whose distaste for dealmaking was one of his most consistent affects—"I would rather face

* Modest indeed—a fraction of what he would eventually spend on rejiggering Watt's patent.

a loaded cannon than settle an account or make a bargain"—managed to drop the hint that his partnership might be subject to improvement, but nothing came of it for more than a year, during which Roebuck's fortunes deteriorated dramatically. In one of the most reliable tropes of his life, Roebuck's talent for finding innovative business opportunities was sabotaged by his chronic inability to make them pay off, and his investment in a coal mine was, literally, underwater. He needed cash, and thought he knew the best way to get it.

In December 1768, at the same moment that Watt's patent application was moving through the London bureaucracy, Roebuck sent a letter to Boulton offering to sell him an exclusive franchise for the Watt engine in three English counties: Warwick, Stafford, and Derby; Boulton declined. In January, Watt, in possession of his first patent, stopped in Birmingham on his way back to Scotland; one can only guess what they discussed, but there can be little doubt that both Watt's plans and Roebuck's offer were shared. In the event, on February 7, 1769, Boulton sent James Watt a letter that read in part:

> I was excited by two motivs [*sic*] to offer you my assistance which were love of you, and love of a money-getting ingenious project. I presum'd that your engine would require mony [*sic*], very accurate workmanship, and extensive correspondence, and the best means . . . of doing the invention justice would be to keep the executive part out of the hands of the multitude of empirical Engineers who from ignorance, want of experience . . . would be very liable to produce bad and inaccurate workmanship. . . . My idea was to settle a manufactory near to my own by the side of our Canal [i.e. in Birmingham] where I could erect all the conveniences necessary for the completion of Engines and from which Manufactory We would serve all the World with Engines of all sizes . . . it would not be worth my while to make for three Countys only, but I find it *very worth while to make for all the World. . . .*" (emphasis added)

James Watt's new engine was a visionary leap—the separate condenser alone doubled the amount of useful work that the Newcomen engine could extract from a given amount of fuel—but its place in history depended on more than Watt's engineering brilliance, perfectionist temperament, or even the grant of a property right to the idea. Watt (and, for that matter, Roebuck) would have been happy to grow prosperous replacing the Newcomen engines at England's coal mines. Changing the world demanded a far larger ambition, and Matthew Boulton was just the man to supply it. It's no coincidence that Boulton's grandiloquent promise to "make for all the world" (one that he would, in the event, redeem), like Albert Einstein's 1939 letter to Franklin Roosevelt warning about possible German development of the atomic bomb, was written in response to a history-shaking example of what is a very nearly universal human phenomenon: the flash of inventive insight.

The nature of which is the subject of chapter 6.

CHAPTER SIX

THE WHOLE THING WAS ARRANGED IN MY MIND

concerning the surprising contents of a Ladies Diary; invention by natural selection; the Flynn Effect; neuronal avalanches; the critical distinction between invention and innovation; and the memory of a stroll on Glasgow Green

It was *in the Green of Glasgow*. I had gone to take a walk on a fine Sabbath afternoon. I had entered the Green by the gate at the foot of Charlotte Street—had passed the old washing-house. I was thinking upon the engine at the time, and had gone as far as the Herd's-house, when *the idea came into my mind, that as steam was an elastic body it would rush into a vacuum, and if a communication was made between the cylinder and an exhausted vessel, it would rush into it, and might be there condensed without cooling the cylinder*. I then saw that I must get quit of the condensed steam and injection water, if I used a jet as in Newcomen's engine. Two ways of doing this occurred to me. First, the water might be run off by a descending pipe, if an offlet could be got at the depth of 35 or 36 feet, and any air might be extracted by a small pump; the second was to make the pump large enough to extract both water and air. . . . *I had not walked farther than the Golf-house when the whole thing was arranged in my mind.*

THE "WHOLE THING" WAS, of course, James Watt's world-historic invention of the separate condenser. It is one of the best recorded, and most repeated, eureka moments since Archimedes leaped out of his bathtub; but accounts of sudden insights have been a regular feature in virtually every history of scientific progress. The fascination with the eureka moment has endured mostly because it turns out to be largely accurate, in general terms if not in detail (no apple actually hit Sir Isaac's cranium, but one falling from a tree in Newton's garden at Woolsthorpe Manor really did inspire the first speculations on the nature of universal gravitation).

Watt's own flash of insight is worth examining not only for its content, but for what it says about insight itself. Those eureka moments are so central to the process of invention that understanding the revolutionary increase in inventive activity demanded by the steam engine also means exploring what modern cognitive science knows (and, more often, suspects) about the mechanism of insight. Watt's moment is just one instance—an earth-shaking one, to be sure—of a phenomenon that is, among humans, nearly as universal as the acquisition of language: solving problems without conscious effort, after effort has failed.

This is not, of course, to say that effort is irrelevant. The real reason that insights seem effortless is that the effort they demand takes place long before the insight appears. It takes a lot of prospecting to find a diamond (to say nothing of the time it took to *make* one), which is why—scrambled metaphors aside—"effortless" insights about musical composition don't occur to nonmusicians. And why, of course, insights about separate condensers don't occur to scholars of ancient Greek. Expertise matters.

This seemingly obvious statement was first tested experimentally in the 1980s by a Swedish émigré psychologist, now at the University of Florida, named K. Anders Ericsson, who has spent the intervening decades developing what has come to be known as the "expert performance" model for human achievement. In study after study of experts in fields as diverse as music, competitive athletics, medicine, and chess, Ericsson and his colleagues were unable to discover any significant inborn difference between the most accomplished performers and the "merely" good. That is, no test for memory, IQ, reaction time, or any other human capacity that might seem to indicate natural talent differentiated the master from the journeyman.

What *did* separate them was, therefore, not inherited, but created; time, not talent, was the critical measurement. Though Ericsson found that both the violinists and basketball players started playing at roughly the same age, the stars in both pursuits spent more time at it than their less accomplished colleagues. Twice as much time, in fact; against all expectations, an expert musician spent, on average, ten thousand hours practicing, as compared to five thousand spent by the not-quite-expert.

The model turned out to apply to a range of pursuits. Cabinetmakers and cardiologists, golfers and gardeners, all became expert after roughly the same amount of time spent mastering their craft. Of all the legends of James Watt's youth, the one no one doubts is that he spent virtually every waking hour of his "apprenticeship" year with John Morgan mastering the skills of fine brasswork, gearing, and instrument repair. His pride in the fine navigational instrument he built as his graduation project is indistinguishable from that felt by a gymnast doing her first back handspring.

James Watt, however, is remembered not as a master clockmaker, but as one of the greatest inventors of all time. And this is where the expert performance model becomes even more relevant. By the 1990s, Ericsson's research was demonstrating that the same phenomenon he had first discovered among concert violinists also applied to the creation of innovations: that the cost of becoming consistently productive at creative inventing is ten thousand hours of practice—five to seven years—just as it is for music, athletics, and chess.

Some of that time is spent acquiring a history of the field: knowledge of what other violinists and inventors have achieved before in order to avoid, in the telling phrase, "reinventing the wheel." The knowledge need not be explicit; the philosopher of science Michael Polanyi* famously thought that leaps of invention were a function of what he called *tacit knowing:* the idea that, in Polanyi's words, "we know more than we can tell." To Polanyi, the acquisition of such internalized knowledge, via *doing,* rather than studying, is necessary preparation of the soil for any true insight.

But knowing that inventors accumulate *knowledge* of other invention doesn't explain how they accumulate *skills* during those ten thousand hours of repetition. Inventing, after all, isn't a craft like basketball, in which mastery is acquired by training muscle and nerve with constant repetition.

Nonetheless, in light of Ericsson's discovery that the route to expert performance looks very similar whether the performance in question is a basketball game, or a chess match, or inventing a new kind of steam engine, it seems worth considering whether the brain's neurons behave like the body's muscles. And, it turns out, they seem to do just that: The more a particular connection between nerve cells is exercised, the stronger it gets. Fifty years ago, a Canadian psychologist named Donald Hebb first tried to put some mathematical rigor behind this well-documented phenomenon, but "Hebbian" learning—the idea that "neurons that fire together, wire together"—was pretty difficult to observe in any nervous system more complicated than that of a marine invertebrate, and even then it was easier to observe than to explain.

In the 1970s, Eric Kandel, a neuroscientist then working at New York University, embarked on a series of experiments that conclusively proved that cognition could be plotted by following a

* A member of the embarrassingly overachieving clan of Hungarian Jews that included Michael's brother, Karl, the economist and author of *The Great Transformation,* a history of the modern market state (one built on "an almost miraculous improvement in the tools of production," i.e., the Industrial Revolution), and his son, John, the 1986 winner of the Nobel Prize in Chemistry.

series of chemical reactions that changed the electrical potential of neurons. Kandel and his colleagues demonstrated that experiences literally change the chemistry of neurons by producing a protein called cyclic Adenosine MonoPhosphate, or cAMP. The cAMP protein, in turn, produces a cascade of chemical changes that either promote or inhibit the synaptic response between neurons; every time the brain calculates the area of a rectangle, or sight-reads a piece of music, or tests an experimental hypothesis, the neurons involved are chemically changed to make it easier to travel the same path again. Kandel's research seems to have identified that repetition forms the chains that Polanyi called tacit knowing, and that James Watt called "the correct modes of reasoning."

Kandel's discovery of the mechanism by which memory is formed and preserved at the cellular level, for which he received the Nobel Prize in Physiology in 2000, was provocative. But because the experiments in question were performed on the fairly simple nervous system of *Aplysia californica*, a giant marine snail, and documented the speed with which the snails could "learn" to eject ink in response to predators, it may be overreaching to say that science knows that the more one practices the violin, or extracts cube roots, the more cAMP is produced. It's even more of a stretch to explain how one learns to sight-read a Chopin etude. Or invent a separate condenser for a steam engine.

Which is why, a decade before Kandel was sticking needles into *Aplysia*, a Caltech neurobiologist named Roger Sperry was working at the other end of the evolutionary scale, performing a series of experiments on a man whose brain had been surgically severed into right and left halves.* The 1962 demonstration of the existence of a two-sided brain, which would win Sperry the Nobel Prize twenty years later, remains a fixture in the world of pop psychology, as anyone who has ever been complimented (or criticized) for right-brained behavior can testify. The notion that

* A remarkable number of discoveries about the function of brain structures have been preceded by an improbable bit of head trauma.

creativity is localized in the right hemisphere of the brain and analytic, linguistic rationality in the left has proved enduringly popular with the general public long after it lost the allegiance of scientists.

However simple the right brain/left brain model, the idea that ideas must originate *somewhere* in the brain's structure continued to attract scientists for years: neurologists, psychologists, Artificial Intelligence researchers. Neuroscientists have even applied the equations of chaos theory to explain how neurons "fire together." John Beggs of Indiana University has shown that the same math used to analyze how sandhills spontaneously collapse, or a stable snowpack turns into an avalanche—the term is "self-organized criticality"—also describes how sudden thoughts, especially insights, appear in the brain. When a single neuron chemically fires its electrical charge, and causes its neighbors to do the same, the random electrical activity that is always present in the human brain can result in a "neuronal avalanche" within the brain.

Where those avalanches ended up, however, remained little more than speculation until there was some way to see what was actually going on inside the brain; so long as the pathway leading to a creative insight remained invisible, theories about them could be proposed, but not tested.

Those new pathways aren't invisible anymore. A cognitive scientist at Northwestern named Mark Jung-Beeman, and one at Drexel named John Kounios, have performed a series of experiments very nicely calibrated to measure heightened activity in portions of the brain when those "eureka" moments strike. In the experiments, subjects were asked to solve a series of puzzles and to report when they solved them by using a systematic strategy versus when the solution came to them by way of a sudden insight. By wiring those subjects up like Christmas trees, they discovered two critical things:

First, when subjects reported solving a puzzle via a sudden flash of insight, an electroencephalograph, which picks up different frequencies of electrical activity, recorded that their brains burst out with the highest of its frequencies: the one that cycles thirty times each second, or 30Hz. This was expected, since this is the frequency band that earlier researchers had associated with similar activities such as recognizing the definition of a word or the outline of a car. What wasn't expected was that the EEG picked up the burst of 30Hz activity three-tenths of a second before a correct "insightful" answer—and did nothing before a wrong one. Second, and even better, simultaneous with the burst of electricity, another machine, the newer-than-new fMRI (functional Magnetic Resonance Imaging) machine showed blood rushing to several sections of the brain's right, "emotional" hemisphere, with the heaviest flow to the same spot—the anterior Superior Temporal Gyrus, or aSTG.

But the discovery that resonates most strongly with James Watt's flash of insight about separating the condensing chamber from the piston is this: Most "normal" brain activity serves to *inhibit* the blood flow to the aSTG. The more active the brain, the more inhibitory, probably for evolutionary reasons: early *Homo sapiens* who spent an inordinate amount of time daydreaming about new ways to start fire were, by definition, spending less time alert to danger, which would have given an overactive aSTG a distinctly negative reproductive value. The brain is evolutionarily hard-wired to do its best daydreaming only when it senses that it is safe to do so—when, in short, it is relaxed. In Kounios's words, "The relaxation phase is crucial. That's why so many insights happen during warm showers." Or during Sunday afternoon walks on Glasgow Green, when the idea of a separate condenser seems to have excited the aSTG in the skull of James Watt. Eureka indeed.

IN 1930, JOSEPH ROSSMAN, who had served for decades as an examiner in the U.S. Patent Office, polled more than seven hundred patentees, producing a remarkable picture of the mind of the inventor. Some of the results were predictable; the three biggest motivators were "love of inventing," "desire to improve," and "financial gain," the ranking for each of which was statistically identical, and each at least twice as important as those appearing down the list, such as "desire to achieve," "prestige," or "altruism" (and certainly not the old saw, "laziness," which was named roughly one-thirtieth as frequently as "financial gain"). A century after *Rocket,* the world of technology had changed immensely: electric power, automobiles, telephones. But the motivations of individual inventors were indistinguishable from those inaugurated by the Industrial Revolution.

Less predictably, Rossman's results demonstrated that the motivation to invent is not typically limited to one invention or industry. Though the most famous inventors are associated in the popular imagination with a single invention—Watt and the separate condenser, Stephenson and *Rocket*—Watt was just as proud of the portable copying machine he invented in 1780 as he was of his steam engine; Stephenson was, in some circles, just as famous for the safety lamp he invented to prevent explosions in coal mines as for his locomotive. Inventors, in Rossman's words, are "recidivists."

In the same vein, Rossman's survey revealed that the greatest obstacle perceived by his patentee universe was not lack of knowledge, legal difficulties, lack of time, or even prejudice against the innovation under consideration. Overwhelmingly, the largest obstacle faced by early twentieth-century inventors (and, almost certainly, their ancestors in the eighteenth century) was "lack of capital." Inventors need investors.

Investors don't always need inventors. Rational investment decisions, as the English economist John Maynard Keynes demonstrated just a few years after Rossman completed his survey, are made by calculating the marginal efficiency of the investment, that is, how much more profit one can expect from putting money into one investment rather than another. When the internal rate of return—Keynes's term—for a given investment is higher than the rate that could be earned somewhere else, it is a smart one; when it is lower, it isn't.

Unfortunately, while any given invention can have a positive IRR, the decision to spend one's life inventing is overwhelmingly negative. Inventors typically forgo more than one-third of their lifetime earnings. Thus, the characteristic stubbornness of inventors throughout history turns out to be fundamentally irrational. Their optimism is by any measure far greater than that found in the general population, with the result that their decision making is, to be charitable, flawed, whether as a result of the classic confirmation bias—the tendency to overvalue data that confirm one's original ideas—or the "sunk-cost" bias, which is another name for throwing good money after bad. Even after reliable colleagues urge them to quit, a third of inventors will continue to invest money, and more than half will continue to invest their time.

A favorite explanation for the seeming contradiction is the work of the Czech émigré economist Joseph Schumpeter,* who drew a famous, though not perfectly clear, boundary between invention and innovation, with the former an economically irrelevant version of the latter. The heroes of Schumpeter's economic analysis were, in consequence, entrepreneurs, who "*may* be inventors just as they may be capitalists . . . they are inventors not by nature of their function, but by coincidence. . . ." To Schumpeter, invention preceded innovation—he characterized the process as embracing three stages: invention, commercialization, and imitation—but was otherwise insignificant. However, his concession that (a) the chances of successful commercialization were improved dramatically when the inventor was involved throughout the process, and (b) the imitation stage looks a lot like invention all over again, since all inventions are to some extent imitative, makes his dichotomy look a little like a chicken-and-egg paradox.

Another study, this one conducted in 1962, compared the results of psychometric tests given to inventors and noninventors (the former defined by behaviors such as application for or receipt of a patent) in similar professions, such as engineers, chemists, architects, psychologists, and science teachers. Some of the results

* In addition to his status as a cheerleader for entrepreneurism—his most famous phrase is undoubtedly the one about the "perennial gale of creative destruction"—Schumpeter was also legendarily hostile to the importance of institutions, particularly laws, and *especially* patent law.

were about what one might expect: inventors are significantly more thing-oriented than people-oriented, more detail-oriented than holistic. They are also likely to come from poorer families than noninventors in the same professions. No surprise there; the eighteenth-century Swiss mathematician Daniel Bernoulli, who coined the term "human capital," explained why innovation has always been a more attractive occupation to have-nots than to haves: not only do small successes seem larger, but they have considerably less to lose.

More interesting, the 1962 study also revealed that independent inventors scored far lower on general intelligence tests than did research scientists, architects, or even graduate students. There's less to this than meets the eye: The intelligence test that was given to the subjects subtracted wrong answers from right answers, and though the inventors consistently got as many answers correct as did the research scientists, they answered far more questions, thereby incurring a ton of deductions. While the study was too small a sample to prove that inventors fear wrong answers less than noninventors, it suggested just that. In the words of the study's authors, "The more inventive an independent inventor is, the more disposed he will be—and this indeed to a marked degree—to try anything that might work."

WATT'S FLASH OF INSIGHT, like those of Newcomen and Savery before him (and thousands more after), was the result of complicated neural activity, operating on a fund of tacit knowledge, in response to both a love of inventing and a love of financial gain. But what gave him the ability to recognize and test that insight was a trained aptitude for mathematics.

The history of mechanical invention in Britain began in a distinctively British manner: with a first generation of craftsmen whose knowledge of machinery was exclusively practical and who were seldom if ever trained in the theory or science behind the levers, escapements, gears, and wheels that they manipulated. These men, however, were followed (not merely paralleled) by another generation of instrument makers, millwrights, and so on, who were.

Beginning in 1704, for example, John Harris, the Vicar of Icklesham in Sussex, published, via subscription, the first volume of the *Lexicon Technicum, or an Universal Dictionary of Arts and Sciences,* the prototype for Enlightenment dictionaries and encyclopedias. Unlike many of the encyclopedias that followed, Harris's work had a decidedly pragmatic bent, containing the most thorough, and most widely read, account of the air pump or Thomas Savery's steam engine. In 1713, a former surveyor and engineer named Henry Beighton, the "first scientific man to study the Newcomen engine," replaced his friend John Tipper as the editor of a journal of calendars, recipes, and medicinal advice called *The Ladies Diary.* His decision to differentiate it from its competitors in a fairly crowded market by including mathematical games and recreations, riddles, and geographical puzzles made it an eighteenth-century version of *Scientific American* and, soon enough, Britain's first and most important mathematical journal. More important, it inaugurated an even more significant expansion of what might be called Britain's mathematically literate population.

Teaching more Britons the intricacies of mathematics would be a giant long-term asset to building an inventive society. Even though uneducated craftsmen had been producing remarkable efficiencies using only rule of thumb—when the great Swiss mathematician Leonhard Euler applied his own considerable talents to calculating the best possible orientation and size for the sails on a Dutch windmill (a brutally complicated bit of engineering, what with the sail pivoting in one plane while rotating in another), he found that carpenters and millwrights had gotten to the same point by trial and error—it took them decades, sometimes centuries, to do so. Giving them the gift of mathematics to do the same work was functionally equivalent to choosing to travel by stagecoach rather than oxcart; you got to the same place, but you got there a lot faster.

Adding experimental rigor to mathematical sophistication accelerated things still more, from stagecoach to—perhaps—*Rocket*. The power of the two in combination, well documented in the work of James Watt, was hugely powerful. But the archetype of mathematical invention in the eighteenth century was not Watt, but John Smeaton, by consensus the most brilliant engineer of his era—a bit like being the most talented painter in sixteenth-century Florence.

SMEATON, UNLIKE MOST OF his generation's innovators, came from a secure middle-class family: his father was an attorney in Leeds, who invited his then sixteen-year-old son into the family firm in 1740. Luckily for the history of engineering, young John found the law less interesting than tinkering, and by 1748 he had moved to London and set up shop as a maker of scientific instruments; five years later, when James Watt arrived in the city seeking to be trained in exactly the same trade, Smeaton was a Fellow of the Royal Society, and had already built his first water mill.

In 1756, he was hired to rebuild the Eddystone Lighthouse, which had burned down the year before; the specification for the sixty-foot-tall structure* required that it be constructed on the Eddystone rocks off the Devonshire coast between high and low tide, and so demanded the invention of a cement—hydraulic lime—that would set even if submerged in water.

The Eddystone Lighthouse was completed in October 1759. That same year, evidently lacking enough occupation to keep himself interested, Smeaton published a paper entitled *An Experimental Enquiry Concerning the Natural Powers of Water and Wind to Turn Mills*. The *Enquiry*, which was rewarded with the Royal Society's oldest and most prestigious prize—the Copley Medal for "outstanding research in any branch of science"—documented Smeaton's nearly seven years' worth of research into the efficiency

* When the original finally wore out, in 1879, a replica, using many of the same granite stones (and Smeaton's innovative marble dowels and dovetails), was rebuilt in Plymouth in honor of Smeaton.

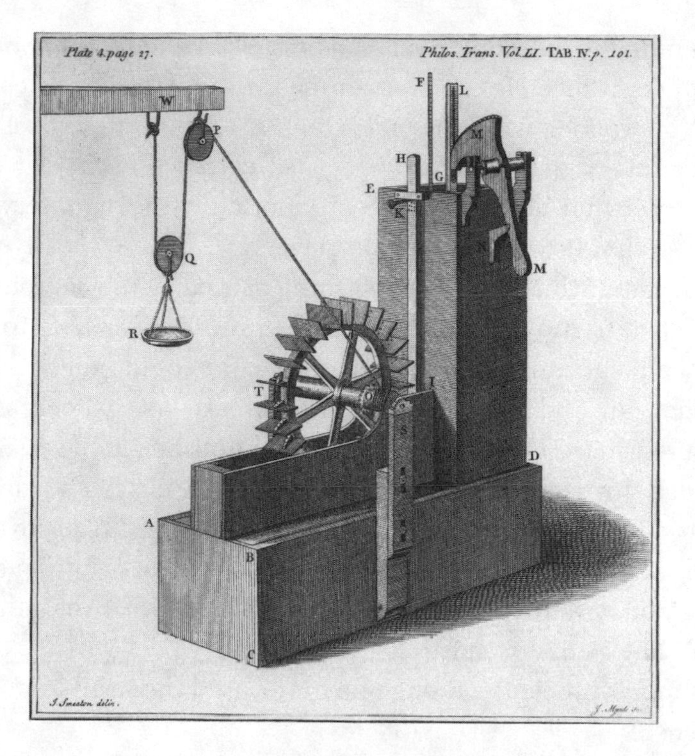

Fig. 4: One of the best-designed experiments of the eighteenth century, Smeaton's waterwheel was able to measure the work produced by water flowing over, under, and past a mill. *Science Museum / Science & Society Picture Library*

of different types of waterwheels, a subject that despite several millennia of practical experience with the technology was still largely a matter of anecdote or, worse, bad theory. In 1704, for example, a French scientist named Antoine Parent had calculated the theoretical benefits of a wheel operated by water flowing past its blades at the lowest point—an "undershot" wheel—against one in which the water fell into buckets just offset from the top of the "overshot" wheel—and got it wrong. Smeaton was a skilled mathematician, but the engineer in him knew that experimental comparison was the only way to answer the question, and, by the way, to demonstrate the best way to generate what was then producing nearly 70 percent of Britain's measured power. His method remains one of the most meticulous experiments of the entire eighteenth century.

He constructed a model waterwheel twenty inches in diameter, running in a model "river" fed by a cistern four feet above the base of the wheel. He then ran a rope through a pulley fifteen feet above the model, with one end attached to the waterwheel's axle and the other to a weight. He was so extraordinarily careful to avoid error that he set the wheel in motion with a counterweight timed so that it would rotate at precisely the same velocity as the flow of water, thus avoiding splashing as well as the confounding element of friction. With this model, Smeaton was able to measure the height to which a constant weight could be lifted by an overshot, an undershot, and even a "breastshot" wheel; and he measured more than just height. His published table of results recorded thirteen categories of data, including cistern height, "virtual head" (the distance water fell into buckets in an overshot wheel), weight of water, and maximum load. The resulting experiment not only disproved Parent's argument for the undershot wheel, but also showed that the overshot wheel was up to two times more "efficient" (though he never used the term in its modern sense).

Smeaton's gifts for engineering weren't, of course, applied only to improving waterpower; an abbreviated list of his achievements include the Calder navigational canal, the Perth Bridge, the Forth & Clyde canal (near the Carron ironworks of John Roebuck, for whom he worked as consultant, building boring mills and furnaces), and Aberdeen harbor. He made dramatic improvements in the original Newcomen design for the steam engine, and enough of a contribution to the Watt separate condenser engine that Watt & Boulton offered him the royalties on one of their installed engines as a thank-you.

But Smeaton's greatest contribution was methodological and, in a critical sense, social. His example showed a generation of other engineers how to approach a problem by systematically varying parameters through experimentation and so improve a technique, or a mechanism, even if they didn't fully grasp the underlying theory. He also explicitly linked the scientific perspective of Isaac Newton with his own world of engineering: "In comparing different experiments, as some fall short, and others exceed the maxim . . . we may, according to the laws of reasoning by induction* conclude the maxim true." More significant than his writings, however, were his readers. Smeaton was, as much as Watt, a hero to the worker bees of the Industrial Revolution. When the first engineering society in the world first met, in 1771 London, Smeaton was sitting at the head of the table, and after his death in 1792, the Society of Civil Engineers—Smeaton's own term, by which he meant not the modern designer of public works, but engineering that was not military—renamed itself the Smeatonian Society. The widespread imitation of Smeaton's systematic technique and professional standards dramatically increased the population of Britons who were competent to evaluate one another's innovations.

The result, in Britain, was not so much a dramatic increase in the number of inventive insights; the example of Watt, and others, provided that. What Smeaton bequeathed to his nation was a process by which those inventions could be experimentally tested, and a large number of engineers who were competent to do so. Their ability to identify the best inventions, and reject the worst, might even have made creative innovation subject to the same forces that cause species to adapt over time: evolution by natural selection.

THE APPLICATION OF THE Darwinian model to everything from dating strategies to cultural history is sometimes dismissed as "secondary" or "pop" Darwinism, to distinguish it from the genuine article, and the habit has become promiscuous. However, this doesn't mean that the Darwinian model is useful only in biology; consider, for example, whether the same sort of circumstances—random variation with selection pressure—preserved the "fittest" of inventions as well.

As far back as the 1960s, the term "blind variation and selective retention" was being used to describe creative innovation without

* This is an explicit reference to Newton's fourth rule of reasoning from Book III of the *Principia Mathematica;* Smeaton was himself something of an astronomer, and entered the Newtonian world through its calculations of celestial motions.

foresight, and advocates for the BVSR model remain so entranced by the potential for mapping creative behavior onto a Darwinian map that they refer to innovations as "ideational mutations." A more modest, and jargon-free, application of Darwinism simply argues that technological progress is proportional to population in the same way as evolutionary change: Unless a population is large enough, the evolutionary changes that occur are not progressive but random, the phenomenon known as *genetic drift*.

It is, needless to say, pretty difficult to identify "progressive change" over time for cognitive abilities like those exhibited by inventors. A brave attempt has been made by James Flynn, the intelligence researcher from New Zealand who first documented, in 1984, what is now known as the Flynn Effect: the phenomenon that the current generation in dozens of different countries scores higher on general intelligence tests than previous generations. Not a little higher: a lot. The bottom 10 percent of today's families are somehow scoring at the same level as the top 10 percent did fifty years ago. The phenomenon is datable to the Industrial Revolution, which exposed an ever larger population to stimulation of their abilities to reason abstractly and concretely simultaneously. The "self-perpetuating feedback loops" (Flynn's term) resulted in the exercise, and therefore the growth, of potential abilities that mattered a lot more to mechanics and artisans than to farmers, or to hunter-gatherers.

Most investigations of the relationship between evolutionary theory and industrialization seem likely to be little more than an entertaining academic parlor game for centuries to come.* One

* The evidence that invention has a Darwinian character is easier to find using the tools of demography than of microbiology, but while the landscape of evolution is large populations, its raw materials are the tiny bits of coded proteins called genes. Bruce Lahn, a geneticist at the University of Chicago, has documented an intriguing discontinuity in the evolutionary history of two genes—microcephalin and abnormal spindle-like microcephaly associated (ASPM)—which, when damaged, are complicit in some fairly onerous genetic disorders affecting intelligence (including big reductions in the size of cerebellums). That history shows substantial changes that can be dated to roughly 37,000 years ago and 5,800 years ago, which are approximately the dates of language acquisition and the discovery of agriculture. This is the first hard evidence that arguably the two biggest social changes in human history are associated with changes in brain size, and presumably function. No such changes dating from the birth of industrialization have been found, or even suspected.

area, however, recalls that the most important inspiration for the original theory of evolution by natural selection was Charles Darwin's observation of evolution by *un*natural selection: the deliberate breeding of animals to reinforce desirable traits, most vividly in Darwin's recognition of the work of pigeon fanciers. Reversing the process, a number of economists have wondered whether it is possible to "breed" inventors: to create circumstances in which more inventive activity occurs (and, by inference, to discover whether those same circumstances obtained in eighteenth-century Britain).

This was one of the many areas that attracted the attention of the Austrian American economist Fritz Machlup, who, forty years ago, approached the question in a slightly different way: Is it possible to expand the inventive work force? Can labor be diverted into the business of invention? Can an educational or training system emphasize invention?

Machlup—who first popularized the idea of a "knowledge economy"—spent decades collecting data on innovation in everything from advertising to typewriter manufacture—by one estimate, on nearly a third of the entire U.S. economy—and concluded with suggesting the counterintuitive possibility that higher rates of compensation actually lower the quality of labor. Machlup argued that the person who prefers to do something other than inventing and does so only under the seductive lure of more money is likely to be less gifted than one who doesn't. This is the "vocation" argument dressed up in econometric equations; at some point, the recruits are going to reduce the average quality of the inventing "army." This is true at *some* point; doctors who cure only for money may be less successful than those who have a true calling. The trick is figuring out *what* point. There will indeed always be amateur inventors (in the original meaning: those who invent out of love), and they may well spend as much time mastering their inventive skills as any professional. But they will also always be fairly thin on the ground compared to the population as a whole.

He also examined the behavior of inventors as an element of what economists call *input-output analysis*. Input-output analysis creates snapshots of entire economies by showing how the output of one economic activity is the input of another: farmers selling wheat to bakers who sell bread to blacksmiths who sell plows back to the farmers. Harvesting, baking, and forging, respectively, are "production functions": the lines on a graph that represent one person adding value and selling it to another. In Machlup's exercise, the supply of inventors (or inventive labor) was the key input; the production function was the transformation of such labor into a commercially useful invention; and the supply of inventions was the output. As always, the equation included a simplifying assumption, and in this case, it was a doozy: that one man's labor is worth roughly the same as another's. This particular assumption gets distorted pretty quickly even in traditional input-output analysis, but it leaps right through the looking glass when applied to the business of inventing, a fact of which Machlup was keenly aware: "a statement that five hours of Mr. Doakes' time [is] equivalent to one hour of Mr. Edison's or two hours of Mr. Bessemer's would sound preposterous."

The invention business is no more immune to the principle of diminishing returns than any other, and in *any* economic system, diminishing returns result anytime a crucial input stays fixed when another one increases. In the case of inventiveness, anytime the store of scientific knowledge (and the number of problems capable of tempting an inventor) isn't increasing, more and more time and resources are required to produce a useful invention. Only the once-in-human-history event known as the Industrial Revolution, because it began the era of continuous invention, had a temporary reprieve from it.

But input-output analysis misses the most important factor of all, which might be called the genius of the system. You only get the one hour of Mr. Edison's time during which he figures out how to make a practical incandescent lightbulb if you also get Mr. Doakes plugging away for five hours at refining the carbonized bamboo filament inside it.

The reason why is actually at the heart of the thing. Mr. Doakes didn't spend those hours because of a simple economic calculus, given the time needed to actually pursue all the variables in all possible frames; Watt's notebooks record months of trying every material under the sun to seal the first boiler of the separate condenser engine. The return on improving even the inventions of antiquity, given the hours, days, and months required and the other demands on the inventor's time, must have been poor indeed. Mr. Doakes spent the time playing the game because he dreamed of winning it.

Which brings us back to James Watt's famous walk on Glasgow Green. The quotation from Watt that opened this chapter appears (not always in the same words) in not only virtually every biography of Watt, but in just about every history of mechanical invention itself, including that of A. P. Usher. Only rarely noted, however, is the fact that Watt's reminiscence first appeared nearly forty years after his death—and was the recollection of two men who heard it from Watt nearly *fifty* years after the famous walk.

Robert and John Hart were two Glasgow engineers and merchants who regarded James Watt with the sort of awe usually reserved for pop musicians, film stars, or star athletes. Or even more: They regarded him "as the greatest and most useful man who ever lived." So when the elderly James Watt entered their shop, sometime in 1813, he was welcomed with adoration, and a barrage of questions about the great events of his life, rather like Michael Jordan beset by fans asking for a play-by-play account of the 1989 NBA playoffs. Watt's recollection of the Sunday stroll down Glasgow Green in 1765 comes entirely from this episode. In short, it is not the sort of memory that a skeptic would regard as completely reliable in all its details.

This is to suggest not that Watt's account is inaccurate, but rather that it says something far more significant about the nature of invention. The research emerging from the fields of information theory and neuroscience on the nature of creative insights offer intriguing ideas about what is happening in an individual inventor's brain at the moment of inspiration. Theories about the aSTG, or cerebellum, or anything else, do not, however, explain much about the notable differences between the nature of invention in the eighteenth century and in the eighth; the structure of the individual brain has not, so far as is known, changed in millennia.

On the other hand, the number of brains producing inventive insights has increased. A lot.

This is why the hero worship of the brothers Hart is more enlightening about the explosion of inventive activity that started in eighteenth-century Britain than their reminiscences. For virtually all of human history, statues had been built to honor kings, soldiers, and religious figures; the Harts lived in the first era that built them to honor builders and inventors. James Watt was an inventor inspired in every way possible, right down to the neurons in his Scottish skull; but he was also, and just as significantly, the inspiration for thousands of other inventors, during his lifetime and beyond. The inscription on the statue of Watt that stood in Westminster Abbey from 1825 until it was moved in 1960 reminded visitors that it was made "Not to perpetuate a name which must endure while the peaceful arts flourish, *but to shew that mankind have learned to know those who best deserve their gratitude*" (emphasis added).

A nation's heroes reveal its ideals, and the Watt memorial carries an impressive weight of symbolism. However, it must be said that the statue, sculpted by Sir Francis Chantrey in marble, might bear that weight more appropriately if it had been made out of the trademark material of the Industrial Revolution: iron.

CHAPTER SEVEN

MASTER OF THEM ALL

concerning differences among Europe's monastic brotherhoods; the unlikely contribution of the brewing of beer to the forging of iron; the geometry of crystals; and an old furnace made new

THE RUINS OF RIEVAULX Abbey sit on a plain surrounded by gently rolling moors not far from the River Wye in the northeast of England. In the years between its founding in 1132 and dissolution in 1536, the abbey's monks farmed more than five thousand acres of productive cropland. In addition to the main building, now a popular tourist stop, Rievaulx included more than seventy outbuildings, spread across a hundred square miles of Yorkshire. Some were granges: satellite farms. Others were cottage factories. And half a dozen were iron foundries, which is why Rievaulx Abbey stands squarely astride one of the half-dozen or so parallel roads that led to the steam revolution, and eventually to *Rocket*. The reason is the monastic brotherhood that founded Rievaulx Abbey, and not at all coincidentally, dominated ironworking (and a dozen other economic activities) in Europe and Britain throughout the medieval period: the Cistercians.

During the eleventh century, the richest and most imitated

monastery in Europe was the Benedictine community at Cluny, in Burgundy. The Cluniacs, like all monastic orders, subscribed, in theory anyway, to the sixth-century Rule of Saint Benedict, an extremely detailed manual for a simple life of prayer and penance. In fact, they were "simple" in much the same way that the Vanderbilt mansions in Newport were "cottages." A Cluniac monk was far likelier to be clothed in silk vestments than in the "woolen cowl for winter and a thin or worn one for summer" prescribed by the Rule. More important for the monastery of Molesme, near Dijon, was the Cluniac tendency to pick and choose pieces of Benedictine doctrine, and to apply more enthusiasm and discipline to their prayers than to their labors.

This was a significant departure from the order's de facto founder, Saint Benedict, who defined labor as one of the highest virtues, and he wasn't referring to the kind of work involved in constructing a clever logical argument. So widespread was his influence that virtually all the technological progress of the medieval period was fueled by monasticism. The monks of St. Victor's Abbey in Paris even included *mechanica*—the skills of an artisan—in their curriculum. In the twelfth century a German Benedictine and metalworker named Theophilus Presbyter wrote an encyclopedia of machinery entitled *Di diversis artibus;* Roger Bacon, the grandfather of experimental science, was a Franciscan, a member of the order founded by Saint Francis of Assisi in part to restore the primacy of humility and hard work.

The Benedictines of Cluny, however, prospered not because of their hard work but because of direct subsidies from secular powers including the kings of France and England and numerous lesser aristocrats. And so, in 1098, the monks of Molesme cleared out, determined to live a purer life by following Benedict's call for *ora et labora:* prayer and (especially) work. The order, now established at "the desert of Cîteaux" (the reference is obscure), whence they took the name "Cistercians," was devoted to the virtues of hard work; and not just hard, but organized. The distinction was the work of one of the order's first leaders, an English monk named Stephen Harding, a remarkably skillful executive who instinctively seemed to have understood how to balance the virtues of flexibility and innovation with those of centralization; by instituting twice-yearly convocations of dozens (later hundreds) of the abbots who ran local Cistercian monasteries all over Europe, he was able to promote regular sharing of what a twenty-first-century management consultant would call "best practices"—in everything from the cultivation of grapes to the cutting of stone—while retaining direct supervision of both process and doctrine. The result was amazing organizational sophistication, a flexible yet disciplined structure that spread from the Elbe to the Atlantic.

Thanks to the administrative genius of Harding and his successors, the order eventually comprised more than eight hundred monasteries, all over Europe, that contained the era's most productive farms, factories—and ironmongeries. Iron was an even larger contributor to the Cistercians' reputation than their expertise in agriculture or machinery, and was a direct consequence of Harding's decision that because some forms of labor were barred to his monastic brothers, others, particularly metalworking, needed to be actively encouraged. The Cistercian monastery in Velehrad (today a part of the Czech Republic) may have been using waterwheels for ironmaking as early as 1269. By 1330, the Cistercians operated at least a dozen smelters and forges in the Burgundy region, of which the largest (and best preserved today) is the one at Fontenay Abbey: more than 150 feet long by nearly thirty feet wide, still bearing the archaeological detritus of substantial iron production.

Which brings us back to Rievaulx. In 1997, a team of archaeologists and geophysicists from the University of Bradford, led by Rob Vernon and Gerry McDonnell, came to north Yorkshire in order to investigate twelfth-century ironmaking techniques. This turns out to be a lot more than traditional pick-and-shovel archaeology; since the earth itself has a fair amount of residual iron (and therefore electrical conductivity), calculating the amount and quality of iron produced at any ruin requires extremely sophisticated high-tech instruments, with intimidating names like magnetometers and fluxgate gradiometers, to separate useful information from the random magnetism found at a suspected ironworking site. What Vernon and McDonnell found caused quite a stir in the world of technological history, which was that the furnaces in use during the thirteenth century at one of Rievaulx Abbey's iron smelters were producing iron at a level of technical sophistication equivalent to that of eighteenth-century Britain. Evidence from the residual magnetism in the slag piles and pits in the nearby village of Laskill revealed that the smelter in use was not only a relatively sophisticated furnace but was, by the standards of the day, huge: built of stone, at least fifteen feet in diameter, and able to produce consistently high-quality iron in enormous quantities. In the line that figured in almost every news account of the expedition, Henry VIII's decision to close the monasteries in 1536 (a consequence of his divorce from Catherine of Aragon and his break with Rome) "delayed the Industrial Revolution by two centuries."

Even if the two-century delay was a journalistic exaggeration—the Cistercians in France, after all, were never suppressed, and the order was famously adept at diffusing techniques throughout all its European abbeys—it deserves attention as a serious thesis about the birth of the world's first sustained era of technological innovation. The value of that thesis, of course, depends on the indispensability of iron to the Industrial Revolution, which at first glance seems self-evident.

First glances, however, are a poor substitute for considered thought. Though the discovery at Laskill is a powerful reminder of the sophistication of medieval technology, the Cistercians' proven ability to produce substantial quantities of high-quality iron not only fails to prove that they were about to ignite an Industrial Revolution when they were suppressed in the early sixteenth century, it actually demonstrates the opposite—and for two reasons. First, the iron of Laskill and Fontenoy was evidence not of industrialization, but of industriousness. The Cistercians owed their factories' efficiency to their disciplined and cheap workforce rather than any technological innovation; there's nothing like a monastic brotherhood that labors twelve hours a day for bread and water to keep costs down. The sixteenth-century monks were still using thirteenth-century technology, and they neither embraced, nor contributed to, the Scientific Revolution of Galileo and Descartes.

The second reason is even more telling: For centuries, the Cistercian monasteries (and other ironmakers; the Cistercians were leaders of medieval iron manufacturing, but they scarcely monopolized it) had been able to supply all the high-quality iron that anyone could use, but all that iron still failed to ignite a technological revolution. Until something happened to increase demand for iron, smelters and forges, like the waterpower that drove them, sounded a lot like one hand clapping. It would sound like nothing else for—what else?—two hundred years.

THE SEVERN RIVER, THE longest in Britain, runs for more than two hundred miles from its source in the Cambrian Mountains of Wales to its mouth at Bristol Channel. The town of Ironbridge in Shropshire is about midway between mouth and source, just south of the intersection with the Tern River. Today the place is home not only to its eponymous bridge—the world's first to be made of iron—but to the Ironbridge Institute, one of the United Kingdom's premier institutions for the study of what is known nationally as "heritage management." The Iron Gorge, where the bridge is located, is one of the United Nations World Heritage Sites, along with the Great Wall of China, Versailles, and the Grand Canyon.* The reason is found in the nearby town of Coalbrookdale.

Men were smelting iron in Coalbrookdale by the middle of the sixteenth century, and probably long before. The oldest surviving furnace at the site is treated as a pretty valuable piece of world

* And, to be fair, the "Struve Geodetic Arc," which consists of thirty-four cairns, obelisks, and rocks-with-holes-drilled-in-them along a fifteen-hundred-mile chain of survey triangulations running from Ukraine to Belarus, and commemorating the nineteenth-century measurement of a longitude meridian. The work of the astronomer Friedrich Georg Wilhelm Struve, it was indeed a noble and memorable achievement, but it draws considerably fewer tourists than, say, Stonehenge.

heritage itself. Housed inside a modern glass pyramid at the Museum of Iron, the "old" furnace, as it is known, is a rough rectangular structure, maybe twenty feet on a side, that looks for all the world like a hypertrophied wood-burning pizza oven. It is built of red bricks still covered with soot that no amount of restoration can remove. When it was excavated, in 1954, the pile of slag hiding it weighed more than fourteen thousand tons, tangible testimony to the century of smelting performed in its hearth beginning in 1709, when it changed the nature of ironmaking forever.

Ironmaking involves a lot more than just digging up a quantity of iron ore and baking it until it's hot enough to melt—though, to be sure, that's a big part of it. Finding the ore is no great challenge; more than 5 percent of the earth's crust is iron, and about one-quarter of the planet's core is a nickel-iron alloy, but it rarely appears in an obligingly pure form. Most of the ores that can be found in nature are oxides: iron plus oxygen, in sixteen different varieties, most commonly hematite and magnetite, with bonus elements like sulfur and phosphorus in varying amounts. To make a material useful for weapons, structures, and so on, those oxides and other impurities must be separated from the iron by smelting, in which the iron ore is heated by a fuel that creates a reducing atmosphere—one that removes the oxides from the ore. The usual fuel is one that contains carbon, because when two carbon atoms are heated in the bottom of the furnace in the presence of oxygen—O_2—they become two molecules of carbon monoxide. The CO in turn reacts with iron oxide as it rises, liberating the oxygen as carbon dioxide—CO_2—and metallic iron.

$$Fe_2O_3 + 3CO \rightarrow 2Fe + 3CO_2$$

There are a lot of other chemical reactions involved, but that's the big one, since the first step in turning iron ore into a bar of iron is getting the oxygen *out;* the second one is putting carbon *in.* And that is a bit more complicated, because the molecular structure of iron—the crystalline shapes into which it forms—changes with heat. At room temperature, and up to about 900°C, iron organizes itself into cubes, with an iron atom at each corner and another in the center of the cube. When it gets hotter than 900°C, the structure changes into a cube with the same eight iron atoms at the corners and another in the center of each face of the cube; at about 1300°C, it changes back to a body-centered crystal. If the transformation takes place in the presence of carbon, carbon atoms take their place in the crystal lattices, increasing the metal's hardness and durability by several orders of magnitude and reducing the malleability of the metal likewise. The percentage of carbon that bonds to iron atoms is the key: If more than 4.5 percent of the final mixture is carbon, the final product is hard, but brittle: good, while molten, for casting, but hard to shape, and far stronger in compression than when twisted or bent. With the carbon percentage less than 0.5 percent, the iron is eminently workable, and becomes the stuff that we call wrought iron. And when the percentage hits the sweet spot of between about 0.5 percent and 1.85 percent, you get steel.

This is slightly more complicated than making soup. The different alloys of carbon and iron, each with different properties, form at different times depending upon the phase transitions between face-centered and body-centered crystalline structures. The timing of those transitions, in turn, vary with temperature, pressure, the presence of other elements, and half a dozen other variables, none of them obvious. Of course, humans were making iron for thousands of years before anyone had anything useful to say about atoms, much less molecular bonds. They were making bronze, from copper and tin, even earlier. During the cultural stage that archaeologists call "the" Iron Age—the definite article is deceptive; Iron Age civilizations appeared in West Africa and Anatolia sometime around 1200 BCE, five hundred years later in northern Europe*—

* The conventional archaeological sequence that leads from stone to bronze to iron "ages" has its critics. The fact that one seems inevitably to follow the other is puzzling on the face of it, given that iron is far easier to find than copper. Many metallurgists have suggested that the fact that primitive iron oxidizes so rapidly may explain the perceived lateness of its arrival.

early civilizations weaned themselves from the equally sturdy bronze (probably because of a shortage of easily mined tin) by using trial and error to combine the ore with heat and another substance, such as limestone (in the jargon of the trade, a *flux*), which melted out impurities such as silicon and sulfur. The earliest iron furnaces were shafts generally about six to eight feet high and about a foot in diameter, in which the burning fuel could get the temperature up to about 1200°C, which was enough for wrought, but not cast, iron.

By the sixteenth century, iron making began to progress beyond folk wisdom and trial and error. The first manuals of metallurgy started to appear in the mid-1500s, most especially *De re metallica* by the German Georg Bauer, writing under the name Agricola, who described the use of the first European blast furnaces, known in German as *Stückofen,* which had hearths roughly five feet long and three feet high, with a foot-deep crucible in the center:

> A certain quantity of iron ore is given to the master [who] throws charcoal into the crucible, and sprinkles over it an iron shovelful of crushed iron ore mixed with unslaked lime. Then he repeatedly throws on charcoal and sprinkles it with ore, and continues until he has slowly built up a heap; it melts when the charcoal has been kindled and the fire violently stimulated by the blast of the bellows. . . .

Agricola's work was so advanced that it remained at the cutting edge of mining and smelting for a century and a half. The furnaces he described replaced the earlier forges, known as bloomeries, which produced a spongelike combination of iron and slag—a "bloom"—from which the slag could be hammered out, leaving a fairly low-carbon iron that could be shaped and worked by smiths, hence *wrought* iron.

Though relatively malleable, early wrought iron wasn't terribly durable; okay for making a door, but not nearly strong enough for a cannon. The *Stückofen,* or its narrower successor, the blast furnace, however, was built to introduce the iron ore and flux at the top of the shaft and to force air at the bottom. The result, once gravity dropped the fuel through the superheated air, which was "blasted" into the chamber and rose via convection, was a furnace that could actually get hot enough to transform the iron. At about 1500°C, the metal undergoes the transition from face-centered to body-centered crystal and back again, absorbing more carbon, making it very hard indeed. This kind of iron—*pig* iron, supposedly named because the relatively narrow channels emerging from the much wider smelter resembled piglets suckling—is so brittle, however, that it is only useful after being poured into forms usually made of loam, or clay.

Those forms could be in the shape of the final iron object, and quite a few useful items could be made from the cast iron so produced. They could also, and even more usefully, be converted into wrought iron by blowing air over heated charcoal and pig iron, which, counterintuitively, simultaneously consumed the carbon in both fuel and iron, "decarbonizing" it to the <1 percent level that permitted shaping as wrought iron (this is known as the "indirect method" for producing wrought iron). The Cistercians had been doing so from about 1300, but they were, in global terms, latecomers; Chinese iron foundries had been using these techniques two thousand years earlier.

Controlling the process that melted, and therefore hardened, iron was an art form, like cooking on a woodstove without a thermostat. It's worth remembering that while recognizably human cultures had been using fire for everything from illumination to space heating to cooking for hundreds of thousands of years, only potters and metalworkers needed to regulate its heat with much precision, and they developed a large empirical body of knowledge about fire millennia before anyone could figure out why a fire burns red at one temperature and white at another. The clues for extracting iron from highly variable ores were partly texture—a taffylike bloom, at the right temperature, might be precisely what the ironmonger wanted—partly color: When the gases in a furnace turned violet, what would be left behind was a pretty pure bit of iron.*

Purity was important: Ironmakers sought just the right mix of iron and carbon, and knew that any contamination by other elements would spoil the iron. Though they were ignorant of the chemical reactions involved, they soon learned that mineral fuels such as pitcoal, or its predecessor, peat, worked poorly, because they introduced impurities, and so, for thousands of years, the fuel of choice was charcoal. The blast furnace at Rievaulx Abbey used charcoal. So did the one at Fontenay Abbey. And, for at least a century, so did the "old" furnace at Coalbrookdale. Unfortunately, that old furnace, like all its contemporaries, needed a *lot* of charcoal: The production of 10,000 tons of iron demanded nearly 100,000 acres of forest, which meant that a single seventeenth-century blast furnace could denude more than four thousand acres each year.

Until 1709, and the arrival of Abraham Darby.

ABRAHAM DARBY WAS BORN in a limestone mining region of the West Midlands, in a village with the memorable name of Wren's Nest. He was descended from barons and earls, though the descent was considerable by the time Abraham was born in 1678. His father, a locksmith and sometime farmer, was at least prosperous enough to stake his son to an apprenticeship working in Birmingham for a "malter"—a roaster and miller of malt for use in beer and whisky. Abraham's master, Jonathan Freeth, like the Darby family, was a member of the Religious Society of Friends. By the time he was an adult, Darby had been educated in a trade and accepted into a religious community, and it is by no means clear which proved the more important in his life—indeed, in the story of industrialization.

* Though, needless to say, early iron smelters wouldn't have known this (or, probably, cared), the violet color was an indication that carbon monoxide is being burned, thus showing the presence of a reducing atmosphere that could remove oxygen from the ore.

Darby's connection with the Society of Friends—the Quakers—proved its worth fairly early. A latecomer to the confessional mosaic of seventeenth-century England, which included (in addition to the established Anglican church) Mennonites, Anabaptists, Presbyterians, Baptists, Puritans, (don't laugh) Muggletonians and Grindletonians, and thousands of very nervous Catholics, the Society of Friends was less than thirty years old when Darby was born and was illegal until passage of the Toleration Act of 1689, one of the many consequences of the arrival of William and Mary (and John Locke) the year before. Darby's Quaker affiliation was to have a number of consequences—the Society's well-known pacifism barred him, for example, from the armaments industry—but the most important was that, like persecuted minorities throughout history, the Quakers took care of their own.

So when Darby moved to Bristol in 1699, after completing his seven years of training with Freeth, he was embraced by the city's small but prosperous Quaker community, which had been established in Bristol since the early 1650s, less than a decade after the movement broke away from the Puritan establishment. The industrious Darby spent three years roasting and milling barley before he decided that brass, not beer, offered the swiftest path to riches, and in 1702, the ambitious twenty-five-year-old joined a number of other Quakers as one of the principals of the Bristol Brass Works Company.

For centuries, brass, the golden alloy of copper and zinc, had been popular all over Britain, first as a purely decorative metal used in tombstones, and then, once the deluge of silver from Spain's New World colonies inundated Europe, as the metal of choice for household utensils and vessels. However, the manufacture of those brass cups and spoons was a near monopoly of the Netherlands, where they had somehow figured out an affordable way of casting them.

The traditional method for casting brass used the same kind of forms used in the manufacture of pig iron: either loam or clay. This was fine for the fairly rough needs of iron tools, but not for kitchenware, which was why the process of fine casting in loam—

time-consuming, painstaking, highly skilled—made it too costly for the mass market. This was why the technique was originally developed for more precious metals, such as bronze. Selling kitchenware to working-class English families was only practicable if the costs could be reduced—and the Dutch had figured out how. If the Bristol Brass Works was to compete with imports, it needed to do the same, and Darby traveled across the channel in 1704 to discover how.

The Dutch secret turned out to be casting in sand rather than loam or clay, and upon his return, Darby sought to perfect what he had learned in Holland, experimenting rigorously with any number of different sands and eventually settling, with the help of another ironworker and Quaker named John Thomas, on a material and process that he patented in 1708. It is by no means insignificant that the wording of the patent explicitly noted that the novelty of Darby's invention was not that it made more, or better, castings, but that it made them at a lower cost: "a new way of casting iron bellied pots and other iron bellied ware in sand only, without loam or clay, by which such iron pots and other ware may be cast fine and with more ease and expedition and may be afforded *cheaper than they can by the way commonly used*" (emphasis added).

Darby realized something else about his new method. If it worked for the relatively rare copper and zinc used to make brass, it might also work for far more abundant, and therefore cheaper, iron. The onetime malter tried to persuade his partners of the merits of his argument, but failed; unfortunately for Bristol, but very good indeed for Coalbrookdale, where Darby moved in 1709, leasing the "old furnace." There, his competitive advantage, in the form of the patent on sand casting for iron, permitted him to succeed beyond expectation. Beyond even the capacity of Coalbrookdale's forests to supply one of the key inputs of ironmaking; within a year, the oak and hazel forests around the Severn were clearcut down to the stumps. Coalbrookdale needed a new fuel.

Abraham Darby wasn't the first to recognize the potential of a charcoal shortage to disrupt iron production. In March 1589,

Queen Elizabeth granted one of those pre–Statute on Monopolies patents to Thomas Proctor and William Peterson, giving them license "to make iron, steel, or lead by using of earth-coal, sea-coal, turf, and peat in the proportion of three parts thereof to one of wood-coal." In 1612, another patent, this one running thirty-one years, was given to an inventor named Simon Sturtevant for the use of "sea-coale or pit-coale" in metalworking; the following year, Sturtevant's exclusive was voided and an ironmaster named John Rovenson was granted the "sole priviledge to make iron . . . with sea-cole, pit-cole, earth-cole, &c."

Darby wasn't even the first in his own family to recognize the need for a new fuel. In 1619, his great-uncle (or, possibly, great-great-uncle; genealogies for the period are vague), Edward Sutton, Baron Dudley paid a license fee to Rovenson for the use of his patent and set to work turning coal plus iron into gold. In 1622, Baron Dudley patented *something*—the grant, which was explicitly exempted when Edward Coke's original Statute on Monopolies took force a year later, recognized that Dudley had discovered "the mystery, art, way, and means, of melting of iron ore, and of making the same into cast works or bars, with sea coals or pit coals in furnaces, with bellows"—but the actual process remained, well, mysterious. Forty years later, in 1665, Baron Dudley's illegitimate son, the unfortunately named Dud Dudley, described, in his self-aggrandizing memoir, *Dud Dudley's Metallum martis,* their success in using pitcoal to make iron. He did not, however, describe how they did it, and the patents of the period are even vaguer than the genealogies. What can be said for certain is that both Dudleys recognized that iron production was limited by the fact that it burned wood far faster than wood can be grown.*

In the event, the younger Dudley continued to cast iron in

* This is the problem with the phrase "renewable energy." Wood is a renewable resource, but for making lumber or paper, not energy. The same adjective can even be applied to fossil fuels like oil and coal, if your time scale for renewability is measured in tens of millions of years. As a good rule of thumb, any resource that captures solar energy in the form of chemical bonds—coal, natural gas, oil, even biofuels—can always be consumed faster than it can be "renewed."

quantities that, by 1630, averaged seven tons a week, but politics started to occupy more of his attention. He served as a royalist officer during the Civil War, thereby backing the losing side; in 1651, while a fugitive under sentence of death for treason, and using the name Dr. Hunt, Dudley spent £700 to build a bloomery. His partners, Sir George Horsey, David Ramsey, and Roger Foulke, however, successfully sued him, using his royalist record against him, taking both the bloomery and what remained of "Dr. Hunt's" money.

Nonetheless, Dudley continued to try to produce high-quality iron with less (or no) charcoal, both alone and with partners. Sometime in the 1670s, he joined forces with a newly made baronet named Clement Clerke, and in 1693 the "Company for Making Iron with Pitcoal" was chartered, using a "work for remelting and casting old Iron with sea cole [*sic*]." The goal, however, remained elusive. Achieving it demanded the sort of ingenuity and "useful and reliable knowledge" acquired as an apprentice and an artisan. In Darby's case, it was a specific and unusual bit of knowledge, dating back to his days roasting malt.

As it turned out, the Shropshire countryside that had been providing the furnace at Coalbrookdale with wood was also rich in pitcoal. No one, however, had used it to smelt iron because, Dudley and Clerke's attempts notwithstanding, the impurities, mostly sulfur, that it caused to become incorporated into the molten iron made for a very brittle, inferior product. For similar reasons, coal is an equally poor fuel choice for roasting barley malt, since while Londoners would—complainingly—breathe sulfurous air from coal-fueled fireplaces, they weren't about to drink beer that tasted like rotten eggs. The answer, as Abraham Darby had every reason to know, was coke.

Coke is what you get when soft, bituminous coal is baked in a very hot oven to draw off most of the contaminants, primarily sulfur. What is left behind is not as pure as charcoal, but is far cleaner than pitcoal, and while it was therefore not perfect for smelting iron, it was a lot cheaper than the rapidly vanishing store of wood. Luckily for Darby, both the ore and the coke available in Shropshire were unusually low in sulfur and therefore minimized the usual problem of contamination that would otherwise have made the resulting iron too brittle.

Using coke offered advantages other than low cost. The problem with using charcoal as a smelting fuel, even when it was abundant, was that a blast furnace needs to keep iron ore and fuel in contact while burning in order to incorporate carbon into the iron's molecular lattice. Charcoal, however, crushes relatively easily, which meant that it couldn't be piled very high in a furnace before it turned to powder under its own weight. This, in turn, put serious limits on the size of any charcoal-fueled blast furnace.

Those limits vanished with Darby's decision in 1710 to use coke, the cakes of which were already compressed by the baking process, in the old furnace at Coalbrookdale. And indeed, the first line in any biography of Abraham Darby will mention the revolutionary development that coke represents in the history of industrialization. But another element of Darby's life helps even more to illuminate the peculiarly English character of the technological explosion that is the subject of this book.

The element remains, unsurprisingly, iron.

IN THE DAYS BEFORE modern quality control, the process of casting iron was highly problematic, since iron ore was as variable as fresh fruit. Its quality depended largely on the other metals bound to it, particularly the quantity of silicon and quality of carbon. Lacking the means to analyze those other elements chemically, iron makers instead categorized by color. Gray iron contains carbon in the form of graphite (the stuff in pencils) and is pretty good as a casting material; the carbon in white iron is combined with other elements (such as sulfur, which makes iron pyrite, or marcasite) that make it far more brittle. The classification of iron, in short, was almost completely empirical. Two men did the decisive work in establishing a scale that was so accurate that it established pretty much the same ten grades used today. One was Abraham Darby; the other was a Frenchman: René Antoine de Réaumur.

Réaumur was a gentleman scientist very much in the mold of Robert Boyle. Like Boyle, he was born into the aristocracy, was a member of the "established"—i.e. Catholic—church, was educated at the finest schools his nation could offer, including the University of Paris, and, again like Boyle at the Royal Society, was one of the first members of the French Académie. His name survives most prominently in the thermometric system he devised in 1730, one that divided the range between freezing and boiling into eighty degrees; the Réaumur scale stayed popular throughout Europe until the nineteenth century, and the incorporation of the Celsius scale into the metric system.* His greatest contribution to metallurgical history was his 1722 insight that the structure of iron was a function of the properties of the other metals with which it was combined, particularly sulfur—an insight he, like Darby, used to classify the various forms of iron.

Unlike Darby, however, he was a scientist before he was an inventor, and long before he was an entrepreneur or even on speaking terms with one. It is instructive that when the government of France, under Louis XV's minister Cardinal de Fleury, made a huge investment in the development of "useful knowledge" (they used the phrase), Réaumur was awarded a huge pension for his discoveries in the grading of iron—and he turned it down because he didn't need it.

Scarcely any two parallel lives do more to demonstrate the differences between eighteenth-century France and Britain: the former a national culture with a powerful affection for pure over applied knowledge, the latter the first nation on earth to give inventors the legally sanctioned right to exploit their ideas. It isn't, of course, that Britain didn't have its own Réaumurs—the Royal Society was full of skilled scientists uninterested in any involvement in commerce—but rather that it also had thousands of men

* Some readers may recall seeing the once ubiquitous advertisements for Charles Minard's map of Napoleon's invasion of Russia, which earned undying fame as, in the words of Edward Tufte, "probably the best statistical graphic ever drawn." The map uses the Réaumur scale to show the temperature confronting the Grande Armée during its retreat from Moscow.

like Darby: an inventor and engineer who cared little about scientific glory but a whole lot about pots and pans.

IF THE CAST IRON used for pots and pans was the most mundane version of the element, the most sublime was steel. As with all iron alloys, carbon is steel's critical component. In its simplest terms, wrought iron has essentially no minimum amount of carbon, just as there is no maximum carbon content for cast iron. As a result, the recipe for either has a substantial fudge factor. Not so with steel. Achieving steel's unique combination of strengths demands a very narrow range of carbon: between 0.25 percent and a bit less than 2 percent. For centuries* this has meant figuring out how to initiate the process whereby carbon insinuates itself into iron's crystalline structure, and how to stop it once it achieves the proper percentage. The techniques used have ranged from the monsoon-driven wind furnaces of south Asia to the quenching and requenching of white-hot iron in water, all of which made steelmaking a boutique business for centuries: good for swords and other edged objects, but not easy to scale up for the production of either a few large pieces or many smaller ones. By the eighteenth century, the most popular method for steelmaking was the cementation process, which stacked a number of bars of wrought iron in a box, bound them together, surrounded them with charcoal, and heated the iron at 1,000° for days, a process that forced some of the carbon into a solid solution with the iron. The resulting high carbon "blister" steel was expensive, frequently excellent, but, since the amount of carbon was wildly variable, inconsistent.

Inconsistently good steel was still better than no steel at all. A swordsman would be more formidable with a weapon made of the "jewel steel" that the Japanese call *tamahagane* than with a more ordinary alloy, but either one is quite capable of dealing mayhem.

* The earliest steel artifacts are more than five thousand years old, but were probably happy accidents of iron manufacture; the earliest reliable steelmaking dates to about the fourth century BCE, in both Asia and the Mediterranean.

Consistency gets more important as precision becomes more valuable, which means that if you had to imagine where consistency in steel manufacturing—uniform strength in tension, for example—mattered most, you could do a lot worse than thinking small. Smaller, even, than kitchenware. Something about the size of, say, a watch spring.

BENJAMIN HUNTSMAN WAS, LIKE Abraham Darby, born into a Quaker farming family that was successful enough to afford an apprenticeship for him with a clockmaker in the Lincolnshire town of Epworth. There, in a process that should by now seem familiar, he spent a seven-year apprenticeship learning a trade so that he was able to open his own shop, in the Yorkshire town of Doncaster, with his own apprentice, by the time he was twenty-one. Two years later, he was not only making and selling his own clocks, but was given the far from honorary duty of caring for the town clock.

In legend, at least, Huntsman entered the history of iron manufacturing sometime in the 1730s out of dissatisfaction with the quality of the steel used to make his clock springs. Since the mainspring of an eighteenth-century clock provided all of its power as it unwound, and any spring provided less drive force as it relaxed, the rate at which it yielded its energy had to be the same for every clock; an even smaller balance spring was needed to get to an "accuracy" of plus or minus ten minutes a day. Given the number of pieces of steel needed to compensate for a machine whose driving force changed with each second, consistency was more highly prized in clockmaking than in any other enterprise.

After nearly ten years of secret experiments* Huntsman finally had his solution: smelting blister steel in the clay crucibles used by local glass makers until it liquefied, which eliminated almost

* Sadly, no records survived Huntsman's fear of industrial espionage, but literally dozens of progressively purer ingots of steel have been discovered where he buried them, testimony to trials at different temperatures, in different environments.

all variability from the end product. So successful was the technique, soon enough known as cast, or crucible, steelmaking, that by 1751 Huntsman had moved to Sheffield, twenty miles north of Doncaster, and hung out a shingle at his own forge. The forge was an advertisement for Huntsman's ingenuity: His crucible used tall chimneys to increase the air draft, and "hard" coke to maintain the cementation temperature. It was also a reminder of the importance of good luck: His furnaces could be made with locally mined sandstone that was strongly resistant to heat, and his crucibles with exceptionally good local clay.

Up until then, Huntsman's life was very much like that of a thousand other innovators of eighteenth-century Britain. He departed from the norm, however, in declining to patent his process, instead relying on the same secrecy he had used in his experiments to be a better protection than legal sanction—supposedly to such a degree that he ran his works only at night. It didn't, however, work. A competitor named Samuel Walker was the first to spy out Huntsman's secret, though not the last; he also attracted industrial spies like the Swede Ludwig Robsahm and Frenchman Gabriel Jars. In 1761 and 1765 respectively they produced reports on the Huntsman process for their own use, and crucible steel became the world's most widely used steelmaking technique until the middle of the nineteenth century.

NEITHER HUNTSMAN'S MAINSPRINGS NOR Darby's iron pots were enough to build *Rocket,* and certainly neither represented enough demand to ignite a revolution. The real demand for iron throughout the mid-1700s was in the form of arms and armor, especially once the worldwide conflict known variously as the Seven Years War or the French and Indian War commenced in 1754. And in Britain, arms and armor meant the navy.

Between the Battle of La Hogue in 1692 and Trafalgar in 1805, the Royal Navy grew from about 250 to 950 ships while nearly doubling the number of guns carried by each ship of the line. Every one of those ships required a huge weight of iron, for everything from the cannon themselves to the hoops around hundreds of barrels, and most of that iron was purchased, despite the advances of Darby and others, from the Bergslagen district of Sweden; in 1750, when Britain consumed 50,000 tons, only 18,000 was produced at home. The reasons were complex, primarily Scandinavia's still abundant forests in close proximity to rich veins of iron ore, but they resulted in a net outflow of £1.5 million annually to Sweden alone. Improving the availability and quality of British iron therefore had both financial and national security implications, which was why, in the 1770s, one of the service's senior purchasing agents was charged with finding a better source for wrought iron. His name was Henry Cort.

FIFTY YEARS AFTER CORT'S death, *The Times* could still laud him as "the father of the iron trade." His origins are a bit murky; he was likely from Lancaster, the son of a brickmaker, and by the 1760s he was working as a clerk to a navy agent: someone charged with collecting the pensions owed to the survivors of naval officers, prize money, and so on. In due course, the Royal Navy promoted him to a position as one of its purchasing agents, where Cort was charged with investigating other options for securing a reliable source of wrought iron.

The choice was simple: either import it from Sweden or make it locally, without charcoal. Darby's coke-fired furnace solved half the problem, but replacing charcoal when turning it into wrought iron—the so-called *fining* process—demanded a new technique. An alternate finery that dates from the 1730s came to be known as the stamp-and-pot system, in which pig iron was cooled, removed from the furnace and "stamped" or broken into small pieces, and then placed in a clay pot that was heated in a coal-fired furnace until the pot broke, after which the iron was reheated and hammered into a relatively pure form of wrought iron.

This limited the quantity of iron to the size of the clay pot, which was unacceptable for naval purposes. Cort's insight was to expand the furnace itself, so that another hearth door opened onto what he called a puddling furnace. In puddling, molten pig iron was stirred by an iron "rabbling" bar, with the fuel separate from the iron in order to remove carbon and other impurities; as the carbon left the pig iron, the melting temperature increased, forcing still more carbon out of the mix. It was a brutal job, requiring men with the strength of Olympic weightlifters, the endurance of Tour de France cyclists, and a willingness to spend ten hours a day in the world's hottest (and filthiest) sauna stirring a pool filled with a sludge of molten iron using a thirty-pound "spoon." Temperatures in the *coolest* parts of the ironworks were typically over 130°; iron was transported by the hundredweight in unsteady wheelbarrows, and the slightest bit of overbalancing meant broken bones. Ingots weighing more than fifty pounds each had to be placed in furnaces at the end of puddler's shovels. Huge furnace doors and grates were regularly opened and closed by chains with a frightening tendency to wrap themselves around arms and legs. In the words of historian David Landes, "The puddlers were the aristocracy of the proletariat, proud, clannish, set apart by sweat and blood. . . . Numerous efforts were made to mechanize the puddling furnace, all of them in vain. Machines could be made to stir the bath, but only the human eye and touch could separate out the solidifying decarburized metal." What the puddlers pulled, taffylike, from the furnace was a "puddle ball" or "loop" of pure iron ready for the next step.

The next step in replacing imported metal with local was forging wrought iron into useful shapes: converting ingots into bars and sheets. Until the late eighteenth century, the conversion was largely a matter of hammering iron that had been softened by heat into bars. A slitting mill run by waterpower then cut the bars into rods, which could then, for example, be either drawn into wire, or—more frequently—cut into nails. At Fontley, near Fareham, seventy miles from London, Cort was the first to invent grooved rollers, through which the softened iron ingots could be converted into bars and sheets, in the manner of a traditional pasta machine. In 1783, he received a patent for "a peculiar method of preparing, welding, and working various sorts of iron."

Cort's title as "the father of the iron trade" was not free and clear, even during his lifetime, since his ingenuity in the improvement of iron purity was not matched by his decidedly impure financing methods. The source of the funds used to purchase the Fontley forge and slitting mills appears to have been money embezzled from the Royal Navy; once this was discovered, Cort lost everything, including his patents.*

Henry Cort's life and works, however, also bear examination by those uninterested in embezzlement, or even metallurgy. Most histories—including, to a degree, this one—that touch on the industrialization of iron production tend to give a lot of attention to men like Darby and Cort while scanting everyone else. This is a bit like assuming that the visible part of the iceberg is the most important part. The portion of an iceberg that floats above the water line is not only a small fraction of the whole, but is "chosen" for its position as much by chance as by any real difference from the portion that remains underwater. The "underwater" story of iron purification is a substantial one: Not only had grooved rollers of a slightly different sort been patented by John Purnell in 1766, but puddling (under a different name) was included in Peter Onions's 1783 patent. Other versions of puddling appeared in William Wood's patent of 1728, the 1763 patent of Watt's partner John Roebuck, the 1776 patent of Thomas and George Cranage (who worked with Darby at Coalbrookdale), John Cockshutt's 1771 patent, and most telling, the four-stage technique that earned John Wright and Richard Jesson patent number 1054 in which the iron was "cleansed of sulphurous matter" inside a rolling barrel.

All of which should serve as a reminder that while the industrialization of Europe was not a function of impersonal demographic forces, neither was it the work of a dozen brilliant

* To the end of his life, Cort contended that he had been the victim of fraud by the Navy Office and other parts of the government; the controversy continues to this day, fueled by charges and countercharges, with as much passion as any debate about the JFK assassination.

geniuses. In any year, the lure of wealth and glory tempted at least a few hundred English inventors, but only a few achieved both.

However, no alloy of copper, or gold, or silver produced anything like the fame and fortune of the iron trade. In his 1910 poem, Rudyard Kipling allegorized the phenomenon:

Gold is for the mistress—silver for the maid;
Copper for the craftsman, cunning at his trade.
"Good!" said the Baron, sitting in his hall,
"But iron—cold iron—is master of them all."

CHAPTER EIGHT

A FIELD THAT IS ENDLESS

concerning the unpredictable consequences of banking crises; a Private Act of Parliament; the folkways of Cornish miners; the difficulties in converting reciprocating into rotational motion; and the largest flour mill in the world

DURING THE FIRST HALF of the eighteenth century, the technology of iron and steel manufacture experienced more innovation than it had during the preceding two thousand years: further evidence of the enthusiasm with which artisans, especially in Britain, greeted the security that both law and culture now granted their inventions. But while the new alchemists—Darby, Huntsman, Cort, and a hundred other, less well remembered inventors—were getting rich by turning base metal into gold, their coke-fueled smelters and puddling furnaces did far more to increase the supply of iron than the demand for it.

This was a bigger hurdle for the inventors of the eighteenth century than for their predecessors, since the incentives that prompted innovations from members of the artisan class were overwhelmingly financial; Leonardo might invent to satisfy his own curiosity, or the ambitions of a Borgia, but Abraham Darby needed customers. So until orders from those customers increased as fast as production, the growth curve of the iron business was at best a straight line with a mild upward slope. And linear growth, even when driven by innovations as dramatic as the atmospheric steam engine, was insufficient, because of population's stubborn tendency to grow exponentially.

Exponential growth in iron production required exponential growth in iron demand. The likely source for that demand, however, wasn't immediately obvious to the era's ironmongers. The Royal Navy, still the largest customer for fabricated iron, was already launched (sorry) on the dramatic buildup that would see its size triple by the time of the Napoleonic Wars, but its ships were still made of wood, and driven by sails. Steam engine components were a promising enough source of new orders for cast and wrought iron that by 1750 the Coalbrookdale foundry alone had built the boilers and cylinders for more than one hundred and fifty Newcomen-style engines. Virtually every one of them, however, used so much fuel to pump water that the only place they were even close to economical was at the heads of the coal mines themselves, which put a severe limit on their appeal.

This was, of course, the opportunity that Matthew Boulton saw in James Watt's separate condenser: An invention that delivered the same amount of pumping force using half the coal would liberate steam engines from the tethers that bound them to their fuel source. And *that* meant that they could be used not only to pump water out of Cornish tin mines, but to pump the bellows of Shropshire iron smelters, and the hammers of Sheffield forges. The Soho Manufactory would buy iron from Britain's foundries and sell them engines in return.

Boulton's February 1769 letter, the one that included his grandiloquent but persuasive offer to make steam engines "for all the world," was tempting to Watt. John Roebuck, on the other hand, was still convinced that the Watt engine was worth a fortune, and that he was the man best equipped to bring it to market. He was correct on the first count, but not the second; throughout Roebuck's life, his business sense was never equal to his vision, and his finances were, in consequence, always in disarray. In September 1769, his need for investment capital became acute enough that he offered to make Boulton a minority partner in his business. They concluded the deal on November 28; Boulton would receive one-third of the rights to the patent in return for £1,000—more than $150,000 in current dollars, about one-third of Roebuck's investment to date; by way of comparison, Watt's 1770 salary as a surveying engineer on the Monkland Canal was £200. With this money in hand, Roebuck was able to finance Watt's continued work at Kinneil, though only fitfully. Three full months in 1770 were taken up in attempting to produce a perfectly round cylinder that would not deform under the pressure of a piston, but the ironmasters at Roebuck's Carron foundries consistently disappointed. Three years after the deal was struck, Boulton's one-third share in the profits from the 1769 patent had precisely the same value as Roebuck's two-thirds: nothing. While Watt was able to support his family surveying canals, and involved himself in other of Roebuck's ventures, the patent wasn't generating enough revenue to buy the coal needed to boil a teapot's worth of water. Roebuck, however, needed those profits. The eighteenth century was no less immune than the twenty-first to credit crises, and Carron's founder was, like thousands of others, caught up in one: In 1772, Glasgow's recently established Bank of Ayr, the country's most liberal in granting credit, collapsed, bringing down virtually every other private banker in Scotland with it. In 1773, Roebuck's business interests, an unsteady edifice at the best of times, were caught up in the panic, and he officially came before the Chancery Court as a bankrupt.

This nullified all of Roebuck's preexisting contracts, leaving a tangle of competing claims, lawsuits, counterclaims, and disputed assets, among them the patent right to James Watt's separate condenser. Roebuck testified that he had spent £3,000 on it, the equivalent of more than sixty man-years of skilled labor, and that another £10,000 (or two hundred man-years) would be needed to make it commercially viable. This seemed a poor bargain to Roebuck's other creditors, "none of whom," in Watt's own words, "value the engine at a farthing," but not to Boulton. He persuaded the other claimants to let him take over 100 percent of the patent rights in return for £630 and an agreement to drop his claims against any other part of the estate. One can imagine them elbowing each other out of the way to get Boulton's signature before he changed his mind.

In December 1774, the last engine made with Roebuck had been dismantled, sent to Birmingham, reassembled, and put to work pumping the water that drove the Soho Manufactory's waterwheels. Watt wrote to his father, "The business I am here about has turned out rather successful."

Watt's comment suggests both a short time horizon and a modest ambition. Boulton had neither. A strategist of both vision and enormous drive, he had acquired the rights to Watt's patent not merely because he saw in it the solution to the Newcomen engine's limits. Britain was thick with purchasers of atmospheric steam engines who had already realized those limits. In Cornwall, for example, copper mines were, in 1775, about to shut down because the coal needed to operate their steam pumping engines was just too expensive. Nor was it the knowledge that the separate condenser had the potential to make the engines affordable for other uses; both Roebuck and Dr. William Small, Watt's original partners, saw the potential, and saw it early. Boulton's genius came rather in his ability to act on this knowledge, both immediately and over the next twenty years.

The short-term objectives for the firm that was not yet called Boulton & Watt were twofold: producing a showpiece engine that would attract the business of Cornwall's mine owners, and securing extension of the existing patent on the separate condenser that would allow the firm to profit from that business for the longest possible time.

So long as Watt was working under Roebuck's supervision at Kinneil, Boulton, as the minority partner, found it impossible to stay informed of the project's successes and failures; in 1770, Watt had attempted to reinforce his balky cylinder with longitudinal supports, but when it failed to keep the cylinder airtight, Roebuck never notified Boulton. With Watt working at Soho, Boulton was able to see the project clearly for the first time, and he calculated that the firm would need until 1800 or so to break even on the investment still needed to perfect the Watt engine.

Unfortunately, the original patent was due to expire in 1783; Roebuck's eagerness to pursue patent protection in 1769 had returned to haunt Watt. Boulton's simple solution was to start over and seek an entirely new patent. In January 1775, Watt, having checked in with his London patent agent, agreed, writing, "we might give up the present patent, and [there is] no doubt that a new one would be granted." Boulton, so inspired, arranged for a bill requesting an Act of Parliament to be introduced in February of 1775, probably with the sponsorship of the Scottish MP William Adam.

It's probably overreaching to call that Act of Parliament, formally the "Fire-Engine Act of 1775," "the most important single event in the Industrial Revolution" (as at least one modern historian termed it), but it does remind the observer that an invention acquires a good bit of its value from the social system in which it is created. It is also a reminder, if another were needed, that even the creators of history's most revolutionary innovations were subject to the same conservative temptation as the most traditional guildsman: to protect advantage once it was acquired. Economists still debate whether the 1775 patent extension promoted or inhibited innovation in steam technology; even the strongest critics concede that if it did retard innovation, it did so for at most a decade, which seems modest enough in the great sweep of history. No one can doubt the importance that Boulton and Watt placed on it. And the effort they were willing to expend in overcoming any obstacles to its passage, including the opposition of the great Whig statesman Edmund Burke, whose enmity Boulton had earned by taking a different side on the question of rights for the fractious—remember, this was 1775—American colonists.

Burke's antagonism was unable to kill the Act, though it did delay its passage until May 10 (the same day Ethan Allen and Benedict Arnold captured Fort Ticonderoga). By then, a combination of clever persuasion and judicious arm twisting had gotten

Matthew Boulton what he wanted: a twenty-five-year extension. The Act reads, in part:

> AND WHEREAS, in order to manufacture these engines with the necessary accuracy, and so that they may be sold at moderate prices, a considerable sum of money must be previously expended in erecting mills, and other apparatus; and as several years, and repeated proofs, will be required before any considerable part of the publick can be fully convinced of the utility of the invention, and of their interest to adopt the same, the whole term granted by the said Letters Patent [i.e. the original 1769 patent] may probably elapse before the said JAMES WATT can receive an adequate advantage to his labor and invention:
>
> AND WHEREAS, by furnishing mechanical powers at much less expense, and in more convenient forms, than has hitherto been done, his engines may be of great utility in facilitating the operations of many great works and manufactures of this kingdom; yet it will not be in the power of the said JAMES WATT to carry his invention into that complete execution which he wishes, and so as to render the same of the highest utility to the publick of which it is capable, unless the term granted by the said LETTERS PATENT be prolonged.

The merits of Boulton's argument aside, the decision to pursue a private Act of Parliament is also a reminder that medieval guildsmen weren't the only ones to see competition the same way that the young Saint Augustine viewed chastity and continence: something to pray for, *sed noli modo:* but not yet. Economists have studied the human predilection for shutting out competitors once they have achieved a position of prosperity, and they refer to the behavior that follows as "rent seeking." Rent-seeking behavior is simply the practice of earning income from an asset without currently working at it. Landlords are rent seekers, particularly to the degree that the land in question is unimproved, and inherited. So are monopolists. And so, indeed, are owners of copyrights and patents, so much so that one of the head-spinning challenges faced by Enlightenment thinkers like Adam Smith was to reconcile their antipathy to all forms of rent seeking, particularly in the form of state-chartered monopolies, with their enthusiasm for inventors and technological innovation. The British Society of Arts, whose mission was "to embolden enterprise, to enlarge science, to refine art, to improve manufacture, and to extend our commerce," granted more than six thousand prizes to successful inventors between 1754 and 1784—but still refused to consider patent-holders for prizes until 1845. Neither Watt nor Boulton was averse to prizes, but they scarcely thought them worth giving up their patents; nor were they averse to work, but they knew that maximizing the value of that work required keeping their competitors from using it.

Legal exclusion of competitors, of course, was valuable only to the degree that there was something valuable to protect, which is why, at the same time that his partner was working on MPs in Westminster, Watt worked furiously on two separate engines in Birmingham. The first was a water pump for the Bloomfield Colliery, a coal mine about fourteen miles from Birmingham, using a cylinder with a diameter of fifty inches, nearly three times the size of the Kinneil model. Watt, and the artisans at the Soho Manufactory—the Birmingham plant then employed six hundred men, with several dozen working directly with Watt—sweated over the new engine for more than a year. They were driven by Boulton's plan to use the machine as an advertisement for Watt's new design, principally the incorporation of a separate condenser, which meant that the engine had to be enormously powerful, highly economical, and utterly reliable—and obviously so.

Boulton got his wish. On March 8, the Bloomfield engine was exhibited to the public amid a ceremony that recalls Thomas Savery's demonstration before the Royal Society at Gresham College seventy-seven years before: the *Birmingham Gazette* recorded "a Number of Scientific Gentlemen whose Curiosity was excited to see the first Movements of so singular and so powerful a Machine, and whose Expectations were fully gratified by the Excellence of its performance. The Workmanship of the Whole did not pass unnoticed, nor unadmired."

But it was the second engine, a 38-incher intended for the blast furnaces at the New Willey ironworks that really deserved all the attention. The Bloomfield colliery engine was still, after all, a water pump, and Boulton knew that pumping water out of mines was only a stepping-stone—a profitable one, to be sure—to something much larger. Boulton wrote, "I rejoice at the well doing of Willey Engine and now hope and flatter my self that we are at the Eve of a fortune," and he was right to rejoice. The engine not only gave the new design an extremely profitable seal of approval and showed off the versatility of the new engine, but it introduced Boulton & Watt to the remarkable figure of John "Iron-Mad" Wilkinson.

WILKINSON WAS A MEMBER of one of the dozen or so families with Nonconformist beliefs (in Wilkinson's case, Presbyterianism) that dominated the iron trade in eighteenth-century England. His father, Isaac, an apprentice foundryman who became a master ironworker, was yet another of the fraternity of onetime artisans who pulled his family into the upper classes by dint of hard work, government contracts, and patent protection. In 1753 he acquired the Bersham Furnace, a foundry built by the family of Abraham Darby just across the Severn from Coalbrookdale, and became a favored supplier of cannon, shot, and shell casings—"engines of mortality of all descriptions"—to the Office of Ordnance, a customer previously neglected by the Quaker Darbys. Four years later, in 1757, he applied for, and received, protection for his design for a cylinder-blowing engine, which used the weight of several columns of water to increase the pressure in a smelter's bellows, and in 1758, a patent for a new method of casting and molding iron.

From the time Isaac's son, John, was sixteen years old, he was a partner in the iron business and in 1757, father and son founded the New Willey Company, a collection of furnaces and smelters on the Severn, about three miles south of Coalbrookdale. Four years later, John was running the entire complex, and doing his best to earn his nickname. He was, indeed, mad for iron; Wilkinson may have appeared grim in person—the profile he had stamped on the copper coins he used as payment in many of his factories practically glowers—but apparently he had a large capacity for love. Not so much for his wife, or his many mistresses; but the man loved his iron. He loved it so much, in fact, that he built an iron pulpit for his church, produced iron writing tablets and pens for his children's school, and, in legend at least, had an iron coffin constructed in which he intended to be interred and that he kept on view in his New Willey office. He even promised to visit his beloved blast furnaces seven years after his death—and his promise was so credible that, on July 14, 1815, thousands of foundrymen showed up expecting to see his ghost. But the *real* business of New Willey, as with Bersham, was the production of cannon.

There are few shapes simpler than a muzzle-loading cannon: an iron tube, closed at one end, with a small hole perpendicular to the long axis. Through that hole, a bit of flame ignited enough gunpowder to send a solid ball of—you guessed it—iron, weighing anywhere from ten ounces to forty-eight pounds, up to a mile and a half downrange. That kind of controlled explosion put a *lot* of stress on even the simplest of tubes, and early cannon had a disconcerting tendency to blow up, often enough because when the tube was cast, imperfections, known as "honeycombs," were introduced that remained invisible until the gunpowder was ignited.

Wilkinson's answer was to cast a solid cylinder of iron and bore out a perfectly circular hole, which created a tighter seal around the cannonball and eliminated the casting imperfections. But drilling a hole in a twelve-foot-long cannon called for a very long drill bit indeed—so long that gravity caused it to deflect downward while drilling.

Wilkinson's great insight was to drill a pilot hole and suspend the boring head at both ends, thus guaranteeing a perfectly round tube, and to rotate the cannon instead of the drill bit. On January

27, 1774, Wilkinson turned his insight into a piece of legally protected property, patent number 1063: "a cylinder attached to a spindle driven by water with the drill stationary except in forward and back motion, as it is mounted on a carriage attached to rack-and-pinion."

The result was phenomenal accuracy, a maximum error estimated at only one-thousandth of an inch. The accuracy was purchased at what seems, from the perspective of the twenty-first century, to be a huge amount of time: In 1800, boring a 64-inch cylinder required twenty-seven working days, and even cannon-sized cylinders could take a week. But New Willey's customers, the Royal Office of Ordnance, were willing to wait, since the alternatives were even worse; without access to Wilkinson's technique, the boring of a forty-inch cylinder for the Philadelphia Water Works took more than four months.

Wilkinson's patent was intended to serve the arts of war more than those of peace, but, like radar, penicillin, and the interstate highway system, it took an unexpected turn. While the finer parts of Watt's engines—the valves, governors, and especially the condenser—could be made by Boulton's craftsmen at Soho, an economical engine needed a large cylinder, made of iron, and that demanded the expertise of others. Darby's Coalbrookdale works, and the Carron Ironworks owned by Roebuck, had been the source for the earlier engines and other experimental versions, but Watt was regularly disappointed with the quality of their work, which he called, with his characteristic perfectionism, "unsound, and totally useless." Even with the earlier innovations in smelting and casting, no one in Britain seemed able to cut the bore of a cylinder in the precise shape of the piston that needed to move within it—ideally with no friction and no leakage of air. Wilkinson's nearly perfect boring system solved this problem.

Boulton and Watt may have encountered Wilkinson as early as 1768, when all three were, briefly and simultaneously, in Birmingham. Certainly they were corresponding not long after the new boring machine was patented; one of the earliest letters to Watt from Wilkinson—who had been hired, in April 1775, to produce the cylinders for both of the demonstration engines—reveals that the ironmaster had as much visionary zeal as Boulton himself: "I wish to do all in the best manner and to start fair. Let us only succeed well in these first engines, particularly in mine, and I will venture to promise you more orders than will be executed in our time. . . . Our time in this world (at best) is but short and we must be busy if you intend that all the engines in this Kingdom shall be put right in our day." For twenty years, beginning in April 1775, when Wilkinson was enlisted to make his first steam engine cylinder for Boulton & Watt, he would remain virtually their sole supplier.

In many ways, Wilkinson had as much invested in the success of the first Boulton & Watt engines as the eponymous firm's owners. He had seen the revolutionary impact of the New Willey engine not only on his own business—it would eventually run not only his bellows but also the forge's stamping hammers and presses—but on all of manufacturing. If the new design caught on, it would create a gigantic new market for symmetrically bored cylinders—a product on which he had, as a result of his 1774 patent, a de facto monopoly.

Boulton was delighted with the ironmonger's enthusiastic embrace of the engine, and shared his belief in the steam-powered factory. However, because he had successfully extended the Watt patent until the year 1800, he had a slightly different timetable, and his plan "to make [engines] for all the world" ran through Cornwall.

CORNWALL, THE SOUTHWESTERN TIP of the British Isles, has been dotted by tin and copper mines since at least Neolithic times. The two components of the alloy that gave the Bronze Age its name were being exported from Cornwall not later than 1000 BCE. When Julius Caesar invaded Britain in 55 BCE, his objectives included Cornish tin and copper.

The men who worked in Cornwall's mines might have thought coal mining in Yorkshire something of a vacation: The sedimentary sheets of coal could be mined with horizontal tunnels off a single mineshaft, while the copper and tin of Cornwall, originally formed from fissures in granite, could only be mined out of individual shafts. By the middle of the eighteenth century, those shafts were the deepest in Britain, as much as eight hundred feet down. As a consequence, the typical Cornish miner traveled to and from his place of work either by whim—placing his foot in a loop of rope lowered by a mule—or by ladder, which was not only dangerous, but exhausting. A contemporary traveler described deep mining in Cornwall thus:

> With hardly room to move their bodies, in sulphureous air, wet to the skin, and buried in the solid rock, these poor devils live and work for a pittance barely sufficient to keep them alive, pecking out the hard ore by the glimmering of a small candle, whose scattered rays will hardly penetrate the thick darkness of the place.

The work was brutal, the business eccentric. In Cornwall, landowners rarely worked the land themselves, instead leasing a "sett" for a period of twenty-one years. The lessors were generally consortia known as "adventurers" ("venture capitalist" is cognate) who put up all costs, and paid the property owner a royalty of between 1/15 and 1/32 of the value of the ore they were able to extract. Adventurers, in turn, appointed "captains" who were responsible for mining operations, including the hiring and management of frequently unruly miners.

Because of this system, Cornish mines were legendary for accounting practices so arcane that, in the words of a correspondent of the *Cornish Telegraph*, "shareholders might grumble over the price of a pennyworth of nails, and pass over without comment the price charged for a steam-engine." The wide range of expertise and interest among those shareholders meant that in what was one of the world's most class-stratified societies, miners, ironmongers, and nobles sat down for dinner four times a year at quarterly "count house dinners."

Inevitably, drainage topped the agenda of these meetings. Because of their depth, draining Cornish mines was absolutely essential, and pumps were almost always needed, in some cases supplemented by drainage tunnels, or adits, as elaborate as the mines themselves. One of them, the Great County Adit, covered more than twenty square miles of Cornish mine country, and in the second half of the eighteenth century, more than half of the most advanced steam engines in the entire world were situated over the Great Adit. As Matthew Boulton realized, the demand for steam pumps, and the distance from coal mines, made Cornwall the perfect laboratory for any invention seeking to reduce the cost of mining.

Boulton's insight would produce a genuinely innovative (not to say genuinely weird) business model. Boulton & Watt took shares in Cornish mines in return for providing engines to drain them, negotiating royalties of one-third of the difference in cost between a Newcomen-type atmospheric engine and a Watt machine doing the same work. This in turn demanded a more precise way of calculating the "same work." The most common calculation of performance was in terms of the so-called "duty," a measurement of the pounds of water raised one foot by a bushel of coal (confusingly, 84 lb. of coal in Newcastle, 88 lb. in London). A high-performing Newcomen-style engine typically generated a duty of between five and nine thousand pounds; that is, a bushel of coal could lift that many pounds of water. A 1778 Watt engine, with separate condenser, achieved a duty of 18,900 pounds.

This was fine for comparing two kinds of steam engines, but less persuasive for customers who were still using wind or animal power to pump water. Watt himself came up with an alternative, measuring the pressure produced by a traditional atmospheric engine, which averaged about seven pounds per square inch, and comparing it with a Boulton & Watt engine using the separate condenser, which generated ten and a half. He then, in his careful way, recorded the effective load lifted by the two engines, using beams of the same length and weight, for the same time period. Finally, he converted it into something he called "horsepower."

The term by then already had a long and fairly inexact history. For millennia, animals (and, all too frequently, other humans) had done most of humanity's work and were the only power sources with variable costs. The outlay for waterpower, which was, by the eighteenth century, even more widely used and easier to measure, was entirely fixed: the construction of a waterwheel. As a result, horses were a favorite way of comparing one sort of pay-as-you-go energy with another. In the first decade of the eighteenth century, Savery himself had promised that his "impellent force" engine could "raise as much water as two Horses working together," and others frequently used horse equivalents as an engineering shorthand. Getting precise about it, however, was a bit more difficult, even when everyone agreed about the need for a standard number. J. T. Desaguliers, the experimental philosopher and lecturer, thought one horse could lift 27,500 pounds a distance of one foot during one minute (or 2,750 pounds five feet in thirty seconds . . . you get the idea); Watt came up with 33,000 or 550 foot-pounds a second (approximately the same as 1,000 pounds at 3 mph for twelve hours a day), which is essentially the same number in use today. Rough though the measure was, horsepower was a relatively effective way to calculate the work done by pumping engines, since the weight and height of the water lifted are the only measurements required.

Watt's horsepower offered a reasonably fair way to balance one engine against another. Boulton's royalty system, however, tilted the balance heavily toward the engine seller, since it obliged buyers to pay one-third the difference for engines based on an ideal annual performance, while the actual engines could be out of commission for months at a time. Even when operating, they frequently weren't *replacing* water or horse wheels, but *supplementing* them, yet intermittent use still demanded constant payment. It was a powerful disincentive against buying a bigger engine appropriate for the future needs of a business, since buyers had to pay a royalty based on an increase in use right away.

Boulton & Watt could afford to ignore any problems the system engendered so long as they were the sole suppliers of the world's most efficient steam engines. In 1795, though, the Birmingham manufacturers discovered that John Wilkinson had been not only providing Boulton & Watt with steam engine parts, but using the same design to sell them on his own, without seeking permission or—far more important—paying a royalty for the privilege. The violent correspondence and litigation that followed is an object lesson in the principle that while ideas may want to be free, their creators very much prefer to keep them leashed.

As a result of the hostility between Wilkinson and his onetime Birmingham customers, they started casting the components of Boulton & Watt engines themselves, building their own ironworks at the Soho Foundry, about a mile from the original Manufactory. But whether the pieces were cast at New Willey or Soho, buyers of Boulton & Watt engines were not getting, except for the cylinder and condenser, a finished engine. In essence, the company sold a kit—no charge for the directions—with "all the cast iron, hammered iron, brass, & copper work . . . including the screw bolts of the cistern, fire door, grate barrs [and] the flywheel with its shaft." They didn't include the boiler, the framing for the engine house, or any of the masonry to hold the contraption, which buyers had to supply themselves, though Boulton & Watt would supply plans and estimates. Even the wooden beam itself that operated all the pumping engines was the responsibility of the buyer, though they did provide engineers to see to the final installation.

It gets better. The royalty charged by Boulton & Watt was based on the difference between the cost of the new engine and the cost of a Newcomen-style pump: the greater the savings, the higher the patent royalties. But even though running the engines was at least as demanding as building them, Boulton & Watt did not supply operators, who might have simultaneously seen to the (usually) smooth operation of the engines and their usefulness to the buyer.

The commercial relationship was further complicated by the usual problems of any industry in its infancy. To cite one critical problem, all the heavy goods needed to be transported from foundry to site at a time when the canal system was barely started—at least one Boulton & Watt engine was too large to travel by canal through the Harecastle tunnel—and *Rocket* was still fifty years in the future, to say nothing of a national rail network. In even shorter supply than transportation were skilled workmen—and they were needed not only to build the "kits" provided by Boulton & Watt (each one custom made, and no two parts assembled anywhere before they arrived on site) but to do so with extreme precision, with the shaft set at an exact right angle to the main beam and perfectly level beneath it and the pistons needing to touch either the top or bottom of the cylinder. It was in response to these demands that Boulton & Watt became what we would today call professional publishers, producing *Directions for Erecting and Working the Newly-Invented Steam Engines* in 1779.

Boulton's plan depended on achieving at least a dozen difficult objectives, each one independent of the other, and each necessary: building a factory that could both cast and machine iron; persuading a group of adventurers to partner with him on a hypothetical cost advantage in fuel (whose future cost couldn't even be predicted with any certainty); training a new workforce to build and operate his engines. Against all odds, it worked. Over the course of five years, Boulton & Watt's steam engines became the pump of choice in the majority of Cornish mines.

Beginning with the installation of an engine at the Ting Tang Mine in November 1776, and the Chacewater Wheal Busy mine a month later, Watt and Boulton succeeded in impressing Cornwall's mine operators, and even the engineers who had been building and servicing Newcomen-style machines for decades. The Chacewater engine, particularly, had been heavily promoted by Boulton as yet another demonstration of the greater efficiency of the new design—he invited hundreds of engineers and mine owners from all over Cornwall to witness the trials, and Watt himself supervised the engine's erection. "All the world are agape," wrote Watt, "to see what it can do," and it gave "universal satisfaction to all beholders, believers or not. . . . Our success here has equaled our most sanguine expectations [and while] our affairs in other parts of England go on very well, no part can or will pay us so well as Cornwall."

Boulton's strategic plan was on schedule. He had discerned an opportunity; had exploited it;* and was soon enough unsatisfied by it. By the end of 1782, he had identified the next conquest for the steam engine, an arena whose potential dwarfed that of the mining industry: wheels.

IT IS NO ACCIDENT that "wheels of industry" is such a cliché description of a manufacturing economy, since the application of force in the form of rotational motion is by far the most important component of useful work. In late eighteenth-century Britain, the wheels that mattered most were the ones turning the mills that ground the nation's grain, and the ones that spun the nation's cloth. Most of them used water; some used wind. None used steam. In December 1782, Boulton wrote to his partner announcing his plan to change that: "I think that these mills represent a field that is endless, and that will be more permanent than these transient mines."

He was scarcely the first to imagine the potential of a factory powered by steam. England's manufacturers had been pestering steam engine makers on the subject for years, hoping to be freed from the shackles that bound them to the flow of rivers or wind. Unfortunately, the machines on offer were piston drivers, and using a piston to run a mill was as sensible as driving a cart by hitting it in the backside with a sledgehammer.

One idea for using a steam-driven piston to produce rotation was nearly a century old: converting linear motion into rotation by taking the new machine and using it to run a familiar one, a waterwheel. Sixty years earlier, Thomas Savery had proposed using the water pumped from mines by the steam engine to oper-

* Not without incurring a lot of resentment among the Cornish miners, who depended on the higher efficiency of the Boulton & Watt engines but hated paying for it—a time bomb that would explode in the 1790s, as we shall see in chapter 9.

ate a 36-foot-diameter waterwheel, housed in a mill house along with the pumping engine. In the 1770s, Boulton & Watt borrowed Savery's idea and recommended using steam engines to pump water not just out of deep mines, but into a reservoir sited well above the engine. Gravity could then pull the water past a waterwheel and so deliver rotational work, and some early steam-powered factories used just such a system, despite its inherent inefficiencies.

And so the challenge of converting the reciprocating motion of the early atmospheric engines into rotary power occupied a fair chunk of the eighteenth century. The fundamental problem of direct conversion was not ignorance; the crank and cam were well known to Newcomen and his successors. In fact, they were known to his predecessors. The teardrop-shaped cam was in use as far back as first-century Greece, and in 1783, John Wilkinson linked a steam engine to his forging hammer by means of a cam and succeeded in raising, and then dropping, the eight-hundred-pound tool. Work that can be transmitted by a cam, however, is the opposite of regular. The transfer of motion from one plane to another—converting the straight-line motion that was intrinsic to any piston-operated engine into the regular power supplied by a rotating wheel—turned out to be as big a challenge as building the steam engine itself. And as important. In the words of the twentieth-century critic Lewis Mumford, "The technical advance which characterizes specifically the modern age is that from reciprocating to rotary motions."

To understand this statement requires (forgive the pun) circling back to the branch of mechanics that describes the transformation of motion from one direction to another, otherwise known as kinematics. In fact, understanding the specific transformation of straight-line into rotary motion requires detours into biomechanics, and even developmental psychology.

IN 1897, THE AUSTRIAN physicist and polymath Ernst Mach (one of the inspirational lights behind the same Gestalt theories so beloved of A. P. Usher) took a vacation from experimenting with the behavior of sound waves to examine the phenomenon of rotary motion. Mach was intrigued by the fact that Neolithic hand querns—really nothing more than two stones between which grain and other substances were ground—used reciprocating, not rotary, motion, with a horizontal handle, for millennia before evolving into rotary querns and finally rotary grindstones, which are really just querns with the upper stone turned ninety degrees. Mach, despite his knowledge of the relatively simple mathematical equations that describe rotary movement, was nonetheless confounded by his observation that infants find rotary motion nearly incomprehensible. In the words of the great medievalist Lynn White, "continuous rotary motion is typical of inorganic matter, whereas reciprocating motion is the sole form of movement found in living things . . . in Europe, at least, crank motion—a kinetic invention more difficult than we can easily conceive—was invented before the crank."

Crank motion was tricky enough; but the first true crank-operated machine, the rotary grindstone, has to be, like the steam engine, one of history's most embarrassingly long-gestating inventions. By the beginning of the third century, cranks were being used in China, typically for winding silk filaments. Arabs were using them as surgical drills for trepanning skulls perhaps six centuries later. But it wasn't until the 1420s that cranks and crankshafts were starting to appear in Europe, in the form of the carpenter's bit and brace; shortly thereafter, the windlass was being used to wind crossbows, and in 1430, an unknown German machinist applied double compound cranks and connecting rods—effectively an extension, or replacement, of the human arm.

The significance of the first combination of the crank with a connecting rod, of course, is that it turned rotary into reciprocal motion, and was therefore the basis of early wind and water power. The earliest visual evidence of a crankshaft connected to a wheel is a drawing by the fifteenth-century painter Pisanello, on display at the Louvre, that shows a water-driven piston pump in front of a crenellated tower, driving two simple cranks and two connecting rods.

Pisanello's pump, however, was never built (though unlike so many of Leonardo's inventions, this one would probably have worked). It wasn't until December 1593 that a Dutch farmer named Cornelis Corneliszoon not only built a wind-powered sawmill, but, because the United Provinces had an even better developed property law than Britain, secured a patent on it lasting fourteen years.* Even better, four years later he extended the patent to embrace the use of a crankshaft that turned the rotational motion of the windmill into the reciprocating motion needed to saw wood; though because it worked in reverse, the linkage used by Corneliszoon's sawmill (which he named *Het Juffertje*, or "The Little Miss"), the first of more than one thousand industrial windmills in the Zaan, is not technically a crank, but a *pitman*.

Reversing the process—getting a piston to drive a wheel rather than the other way around—is a trivial enough task. Attach the crank (or crankshaft: a rod connected to a wheel by a right-angled arm that pivots with the rotation of the wheel) or cam directly to the piston. But getting that wheel to rotate at a constant speed was a different matter, and any useful mechanical solution required something that smoothed out the variability in the push-pull motion of a piston—something that eliminated the "dead spot" when the piston reversed direction.

Rotating disks had been used to smooth out those dead spots—to store kinetic energy—since Neolithic times. Several millennia later, the equations of Kepler and Newton explained why: Angular momentum tends to be conserved. This is the same phenomenon that makes an ice skater spin faster when her arms are drawn in to her body, since her angular momentum is the product of her moment of inertia and angular velocity; bringing her arms closer to her axis of rotation reduces her moment, which means her velocity must increase.

A disk rotating around an axle—better known as a flywheel—does the same trick in reverse, keeping speed constant when

* Better developed, but less successful. See chapter 11.

power input is intermittent, as it always is with a piston or other reciprocating motion. Potter's wheels, an early example, take the pumping motion of a hand or foot and convert it into relatively smooth rotation. The first flywheel used as part of a larger machine appears in Theophilus Presbyter's *Di divers artibus;* by the fourteenth century, the Parisian philosopher John Buridan was observing that rotary grindstones store power, because they keep turning even when no one is turning them.

It is baffling, except perhaps in light of some infantile resistance to mixing rotary with reciprocal motion, that combining the crank with the flywheel took centuries rather than months. Not until 1740 did an inventor named John Wise suggest using a flywheel—he called it a "double tumbling wheel" in his patent*—to produce rotary motion from a Newcomen-type engine, and *that* went nowhere until 1778, when the engineer and inventor Matthew Wasbrough, originally from Bristol but working in Birmingham, took the first step toward a useful rotary engine, patenting (in March 1779, number 1213—things were accelerating) a complicated assortment of pulleys, wheel segments, and flywheels ("to render the motion more regular and uniform").

Two years later, Wasbrough incorporated a crank and flywheel into an iron rolling mill, but the circumstances were tainted. Watt, and his assistants at the Soho Manufactory, had been preparing a new design that would produce rotary motion from their existing—and highly successful—piston engine. They had been working on their own crankshaft model for more than a year when a workman at Soho leaked the plans to Wasbrough and his partners, John Steed and James Pickard. In August 1780, the three secured a patent (number 1263) for a beam-operated engine using the crank plus connecting rod (though how connected, the patent does not say).

Watt had not believed the crank plus rod novel enough to be patentable, because of its long history. This belief didn't do much

* Wise's patent number was 540—forty-two years after Thomas Savery received number 356 for his "fire engine," which adds up to fewer than five patents a year.

to calm him. In most aspects of his life, James Watt was a gentle sort, but he was extraordinarily sensitive when it came to his inventions, and he was sputteringly furious at what he saw as Wasbrough's treachery, writing to Boulton, "I know the contrivance is my own, and has been stolen from me by the most infamous means. . . . Had I esteemed [Wasbrough] a man of Ingenuity and the real inventor of the thing in question, I should not have made any objection, but when I know the contrivance is my own and has been stole [*sic*] from me by the most infamous means and to add to the provocation a patent surreptitiously obtained for it, I think it would be descending below the Character of a Man to be any ways aiding or assisting to him or to his pretended inventions."

Watt's pride in his inventions may have been the defining characteristic of his personality. It certainly explains the amount of time he would spend in subsequent years defending his own patent claims, and frequently the claims of others, in court. The immediate reaction, however, was to take his revenge not in a courtroom, but in his workshop, and not working with a lawyer, but with another engineer, the remarkable William Murdock.

MURDOCK—ORIGINALLY MURDOCH; the Scottish spelling was anglicized—was only twenty-two years old in 1777 when he left Ayrshire in Scotland and walked three hundred miles to Birmingham, hoping for employment at Boulton & Watt. In legend, at least, Boulton asked about the young Scot's strange-looking hat, and when he answered that the hat was made of "timmer [timber] . . . turned on my little lathey [lathe]," he evidently decided that the young Scot's mechanical skills were sufficient to overcome his unfortunate accent. Two years later, his employers trusted him enough that they assigned him to build, unsupervised, what was only their fourth engine. Before the year was out, he was the firm's supervisor in Cornwall.

Murdock's relationship with Watt was complicated. The nearly twenty-year difference in their ages, their mutual affection and loyalty, their status as employee and employer, their shared Scottish heritage, and their technical brilliance in the same disciplines made for one of the more fraught Oedipal conflicts of the era. Given all the potential arenas for conflict, it is actually somewhat surprising that the relationship survived as long as it did, with the two apparently agreeing to annoy one another mercilessly in lieu of more bloody combat. Watt, in particular, regularly complained about Murdock's tendency to second-guess (and, even more annoyingly, to improve) the master's ideas. Anyone who has ever supervised a talented subordinate with a tendency to set his own priorities will find Watt's letters familiar: "I wish William could be brought to do as we do, to mind the business in hand, and let such as Symington [William Symington, the builder of the *Charlotte Dundas,* one of the world's first steam-engine boats] and Sadler [James Sadler, balloonist and inventor of a table steam engine] throw away their time and money, hunting shadows."

Even so, Watt was nothing if not fair-minded. He may have resented the time Murdock spent inventing rather than on the company's real business, which was installing, and collecting royalties based on the continued performance of, the Watt engines.* He was, however, just as often delighted by the inventions he made on behalf of Boulton & Watt, including a compressed air pump and a cement that would bond two pieces of iron, calling Murdock "the most active man and best engine erector I ever saw." His value was never, however, to be higher than when Watt enlisted his talents on what came to be known as the sun-and-planet gear.

If the crank and its cousins are methods for turning reciprocation into rotation, then gears, broadly speaking, are methods for turning one form of rotation into another. The primary reason anyone would want to perform such a trick is the phenomenon known as the *mechanical advantage,* which is the formal term for the fact that when the teeth of a larger gear engage with those of a

* Watt was not exactly immune to the temptation. During the same period, he patented a dozen other unrelated inventions, including a steam press for linen and muslin cloth.

smaller, it must rotate faster for each shared revolution; since torque is defined as circumferential force multiplied by the radius, the bigger the radius, the greater the torque.

The theory and practice of gears were well known to the Greeks of the first century BCE, who were not only familiar with several basic forms of gearing (including the familiar version, in which the teeth project radially from the gear's center, and the worm, in which another Greek invention, the screw, was set at right angles to a traditional gear) but were also capable of incorporating them into geared engines—though the nature of those engines is revealing. The best-known surviving example, the so-called Antikythera mechanism (so called because Greek sponge fishermen discovered it off the eponymous island in 1900), contains at least eighty separate bronze gears in a wooden case not much larger than the laptop computer on which this is being written.

Antikythera's elaborately linked gears, probably used to calculate the positions of planets and stars on a particular date, posed an unanswered question to scholars for decades: Why was all this extraordinary precision and technological expertise—no comparable mechanism would appear until the clocks of the mid-eighteenth century—not used to produce anything that was recognizably useful?

The answer is related to the similar question that might be asked about Hero of Alexandria: Antikythera's creators were the ancestors of toymakers (or, at best, clockmakers), not engineers.

Technical skill is artifact-neutral, and is just as likely to be applied to toys as tools. The builders of the Antikythera gears, like Hero, or even the great clockmakers of the Middle Ages, were just as mechanically adept as Watt or Newcomen, but they responded to a different set of incentives. Those incentives were the satisfaction of an elite segment of society with the means to commission astonishingly intricate clocks and calculators or pyramids and cathedrals. By their very nature, devices like Antikythera were built to satisfy a very different set of demands than those of a mill owner shopping for a steam engine.

Eighteen centuries later, inventors were responding to a new set of incentives—which might be the best way of explaining the

Industrial Revolution. The laws of supply and demand created the market for steam-powered mills. The laws of mechanics limited the number of ways in which the back-and-forth movement of a piston could be translated into the smooth rotation of a wheel. And Britain's property laws excluded the best one: the crankshaft. Prohibited from copying Wasbrough's crank, Watt and Murdock nonetheless had to compete with it.

And not just compete—preempt. Whether or not Pickard and Wasbrough were guilty of theft, they had certainly convinced Watt that he was a victim of it, and he was determined not to be one again. So determined, in fact, that the patent application he and Murdock submitted in 1781 included five separate inventions, including a "swashplate" that used the engine to raise a beam through an arc that looks like nothing so much as an amusement park ride, like the pirate ship that swings high enough for its passengers to go from nearly vertical to horizontal to vertical again; a counterweighted crank wheel, with the wheel divided along a diameter with one half heavily unbalanced; and an "eccentric" wheel inside an external yoke. Most important was Murdock's sun-and-planet gear, which linked a connecting rod to one gear that made an orbit around another, larger gear like a planet circling the sun. The last one not only smoothed out the power curve, but "gave the additional advantage that the output shaft rotated at twice engine speed."*

The application, for patent number 1306, read in part, "for certain new methods of applying the vibrating or reciprocating action of steam or fire engines, to produce a continued rotative or circular motion round an axis or centre, and thereby to give motion to the wheels or mills or other machines."† It was sub-

* The doubling of speed was useful everywhere, but would be a huge advantage in textile mills, since spinning yarn demanded regular increases in the speed of rotation—as will be seen in chapter 10.

† A book entitled *The Diverse and Artifactitious Machines of Captain Agostino Ramelli* by a sixteenth-century Italian inventor in the employ of the Medicis contained illustrations that, whether known to Watt and Murdock or not, seem to prefigure the sun-and-planet gear.

mitted on October 25, 1781, but the law allowed four months to create the final specifications. Watt needed every day. Predictably, his perfectionism was once again his greatest asset and biggest liability; he regularly complained about both his health, and the quality of his assistants' work; to Boulton he wrote, "I wish you could supply me with a draughtsman of abilities [as] I tremble at the thought of making a complete set of drawings. . . . I must drag on a miserable existence the best way I can" even though afflicted with "backache, headache, and lowness of spirits. . . ."

But he also produced his most important patent specification

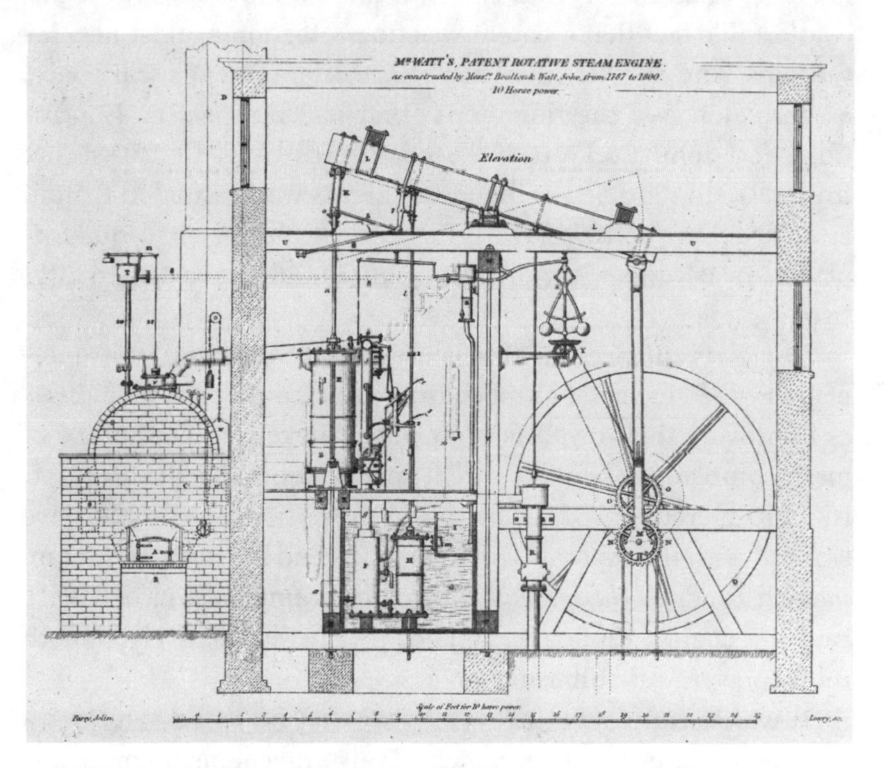

Fig. 5: The caption for this technical drawing reads "Mr. Watt's Patent Rotative Steam Engine as constructed by Messrs. Boulton & Watt, Soho, from 1787 to 1800. 10 Horse power." By 1787, the engine had evolved considerably from the earlier versions, using the sun-and-planet gear to drive the large wheel; the Watt linkage to connect the beam with the cylinder, on the left; and even Watt's feedback-driven flyball governor—the two balls hanging above and to the left of the large wheel—to control the wheel's speed. *Science Museum / Science & Society Picture Library*

since the separate condensing grant of 1769, "the neatest drawing I had ever made."

THE SUN-AND-PLANET (or, for that matter, the crank plus connecting rod, which was, after all, Watt's first choice for producing rotary motion, and would be everybody's after the Wasbrough patent expired in 1794) was a huge step toward the introduction of steam power into mills and factories, rather than pumps. But it was only a step. The lesson of the Wasbrough imbroglio was not merely that Boulton & Watt needed to improve security at the Soho Manufactory, but also that they had to confront the uncomfortable fact that a patent was no protection against new inventions. And by 1781, invention was accelerating at a scary pace, a consequence of the time bomb that had been set by Edward Coke and John Locke in the century preceding. Consider that from 1700 to 1740, fewer than five patents were issued in Britain annually; from 1740 to 1780, the annual number had quadrupled, to nearly nineteen, and from 1780 to 1800, it was up to fifty-two.

One consequence was that successful technological entrepreneurs needed to spend as much time anticipating new inventions as improving their own. As Watt put it (in responding to one of many complaints about his patent specification), had he been satisfied to merely introduce his initial inventions, he would have stopped with the separate condenser; instead his mind ran "upon making Engines *cheap,* as well as *good*" (emphasis in original). And the engine, though capable of producing relatively smooth rotary power, was still using too much coal.

It was not until the following year that he had a solution: a double-acting cylinder, in which a valve mechanism connected each end of the cylinder with both the condenser and the boiler. When the top was connected with the boiler, the bottom was connected with the condenser, and vice versa. The result was that the upward and downward strokes were identical. When the piston was up, the steam in the portion of the cylinder beneath it was condensed, causing a vacuum. And when it was down, the upper portion of the cylinder was likewise condensed. The piston was therefore both pushed and pulled, both up and down. Watt measured the pressure drop in his own cylinders, and used Boyle's Law—the one that demonstrated the relationship between gas pressure and volume—to calculate that he could cut fuel consumption in the engine by three-quarters, while losing only one-half of the power.

In March of 1782 Watt patented the double-acting engine. The increase in power output per measure of coal was dramatic; the incorporation of a flywheel made the power even smoother. The smooth rotation of a steam-operated flywheel is one of the reasons for the rather eerie quiet of even large steam engines, at least as compared to their internal combustion counterparts; conversations in normal tone of voice can easily be heard in front of the 1,200-hp Brooklands Mill engine, built in 1893, which uses a flywheel thirty feet in diameter; a 1795 description of the first flywheel-operated steam engines remarked that they are "scarce heard in the building where they are erected." But the double-acting engine also introduced a new problem: The piston needed to move the connecting rod precisely up and down.

Like so many things about steam engines, this sounds trivial but isn't, since it illustrates what is one of the most distinctive characteristics of the steam revolution: Inventions don't just solve problems; they create new ones, which demand—and inspire—other inventions. The double-acting engine represented a large leap in power output for every bushel of coal, but its piston now needed not just to pull a beam down but to push it up. To do so most efficiently, that piston needed to travel along a straight vertical line, or at least as straight a line as possible. However, the beam to which it was connected didn't travel straight up and down, but along an arc, which meant its angle was constantly changing, pushing the piston off line. Earlier engines, with their single power stroke, could use a chain to pull the beam down, but obviously chains could not transmit force in two directions and wouldn't stay straight in either direction.

Which is why Watt took out another patent, in 1784, "for certain new improvements upon fire and steam engines and upon Machines worked or moved by the same."* The new patent included the Watt linkage, a pair of horizontal rods mounted parallel to one another, connected by a pivot to a third, perpendicular rod, which kept the piston in a straight line and so reduced the friction that caused wear on the inside of the cylinder, and, more important, wasted energy in doing so. Watt himself was uncharacteristically enthusiastic about it; in a letter to Matthew Boulton of June 1784, he announced "one of the most ingenious, simple pieces of mechanism I have invented." In 1808, he wrote, "I am more proud of the parallel motion than of any other mechanical invention I have ever made."

His pride was understandable. The double-acting engine, transformed by the use of flywheels, parallel linkages, and the sun-and-planet gear, liberated steam power from Britain's mines, and Britain's mills from the banks of her rivers. Steam engines could now be built anywhere that either reciprocating or rotary motion was needed. This was decisively proved at a site on the south side of the Thames, in London itself.

Matthew Boulton, who understood not only the potential for grain mills run by steam engines but also the need for promoting that vision to a less enlightened public, first broached the idea of a steam-powered flour mill in London sometime in 1783. While Britain's representatives were in Paris signing a peace treaty with their former American subjects, Boulton was having somewhat less luck hammering out agreements with both the investors needed to finance his showpiece factory and the local millers whose approval was a prerequisite to a peaceful operation. Eventually failing on both counts (in a letter of 1784, he demolished each of their arguments, including the concern that the proposed mill would produce too much coal smoke, figuratively throwing

* It also included a description of a piston-driven steam carriage that used "the elastic force of steam to give motion." Watt had no intention of building one, for reasons that will become apparent, but he had already mastered the technique of using patents to preempt his competition.

up his hands with the observation that "the millers are determined to be masters of us and the public"), he decided to proceed anyway.

The plans for the new mill were drawn up, at Boulton's direction, by one of Britain's most influential architects, Samuel Wyatt, to be built on a site near the south end of the original Blackfriars Bridge. Fans of odd coincidences might take some pleasure in the fact that the north end of the bridge abutted the same Temple Bar where Edward Coke had drafted the original Statute of Monopolies, the law under which the separate condenser, sun-and-planet gear, and parallel linkage were protected. Or the slightly less coincidental fact that the engines at the mill were erected by a Boulton & Watt employee named John Rennie—the same man who would, twenty years later, build engine 42B at the Crofton pumping station on the Kennet and Avon Canal.

The Albion Mills, as it would be called, was built on a scale hitherto unimagined. The largest flour mill in London in 1783 used four pairs of grinding stones; Albion was to have thirty, driven by three steam engines, each with a 34-inch cylinder. Within months after its completion, in 1786, those engines were driving mills that produced six thousand bushels of flour every week—which both fed a lot of Londoners and angered a lot of millers.

The Albion Mills was London's first factory, and its first great symbol of industrialization; its construction inaugurated not only the great age of steam-driven factories,* but also the doomed though poignant resistance to them. That resistance took the shape of direct action—no one knows how the fire that destroyed the Albion Mills in 1791 began, but arson by millers threatened by its success seems likely—and even poetry. It was, after all, the

* It also inspired one of Watt's best-remembered inventions. In 1789, improving on an earlier device created for grain mills like Albion, he invented a centrifugal governor: two metal flyballs held in orbit around a vertical pole, linked to the power stroke of the steam engine. As the engine speeded up, centrifugal force lifted the "arms" to which the balls were connected; with a greater distance to travel with each rotation, the balls spun faster, but as they did so, they pulled down rods mounted on top of the balls and attached to the top of the pole, which in turn closed the throttle, thus slowing the engine.

blackened ruins of Albion Mills that inspired Lambeth resident William Blake to write, in "Jerusalem,"

And did the Countenance Divine
Shine forth upon our clouded hills?
And was Jerusalem builded here
Among these dark Satanic Mills?

Blake's vision remains powerful and chilling, but he was still whistling past a graveyard; it was the factory, in the end, that was to triumph. Albion Mills stood for only five years, but its proof of the ability of the steam engine to produce rotary power anywhere it was needed was decisive. Behind the Albion Mills engine were hundreds of large and small innovations that had solved a dozen ancient problems in physics, metallurgy, and kinematics. Before it was a new one: how to achieve clocklike (or Antikythera-like) precision in an industrial machine.

CHAPTER NINE

QUITE SPLENDID WITH A FILE

concerning the picking of locks; the use of wood in the making of iron, and iron in the making of wood; the very great importance of very small errors; blocks of all shapes and sizes; and the tool known as "the Lord Chancellor"

THE SHOP WAS SLIGHTLY off the most traveled portion of Piccadilly, across from Green Park near Half Moon Street, but hundreds, if not thousands, of pedestrians still walked past it every day. Which meant that every day during the spring of 1801, thousands of eyes saw the challenge. It was hung in the store's window, incised on a brass contraption the shape of an oversized acorn, and read:

THE ARTIST
WHO CAN MAKE AN
INSTRUMENT THAT WILL PICK
OR OPEN THIS LOCK SHALL
RECEIVE 200 GUINEAS
THE MOMENT
IT IS PRODUCED
APPLICATIONS IN WRITING ONLY

At the top of the acorn, in place of the cupule that attaches the nut to its tree, was the lock in question. At its bottom was the legend:

CAUTION.
THE PUBLIC IS RESPECTFULLY INFORMED
THAT EVERY LOCK MADE BY
BRAMAH & C° IS STAMPED
WITH THEIR ADDRESS
124 PICCADILLY

As gimmicks go, this was one of the more successful in the history of marketing.* But its object was no gimmick: the "challenge lock," designed to be opened by a tubular key incised with slots along the long axis, would defy all challenges for fifty years. Different models were on offer almost from the beginning, containing as few as five sliders or as many as fourteen. That was the number contained in the "challenge lock," giving it potentially 470 *million* possible combinations and making it virtually unpickable. Indeed, it remained unpicked until 1851, when an American locksmith finally succeeded at the Great London Exposition, supposedly taking more than fifty hours in the process. The challenge lock was such a remarkable bit of technological brilliance that an updated version of the same design is still sold by the same firm, now known as Bramah UK. Its managing director, Jeremy Bramah, is a direct descendant of the firm's founder, Joseph Bramah, and it is no coincidence that his original lock is

* Estimating the present-day value of the amount in question—of any monetary amount from the period—is problematic. The historical purchasing power of the pound sterling can be calculated either by compounding the changes in the retail price index or as a fraction of average earnings in Britain. Using the first method, the prize was worth a little less than £12,000 in 2009 currency; the second, more than £160,000. Put another way, the cost of a loaf of bread has increased by about sixty times; but the hour that a laborer had to work to earn the money to buy that loaf is now only five minutes. The huge discrepancy between the two calculations is in itself a powerful reminder of the transformative power of industrialization.

on display at London's Science Museum only yards away from *Rocket.*

JOSEPH BRAMAH, THE SON of a Yorkshire farmer, had been apprenticed to a cabinetmaker named Joseph Allott in 1765 when he was sixteen years old, but it took another thirteen years for him to get his first patent, for a flush toilet with a floating ball and flap valve designed to prevent it from seizing up during freezing weather. It was not only an immediate success, with thousands installed across Britain, but is still recognizably the system used in most modern water closets.

Bramah may have been a late bloomer, but during the last twenty years of his life he would become an inventing phenomenon, creating dozens of highly profitable machines for the widest possible range of applications. He was also, from 1783 on, a member of the British Society of Arts, which, it will be recalled, was enthusiastic about giving prizes for inventions but opposed patenting them until the middle of the nineteenth century. One can only imagine the debates that must have been prompted by Bramah's presence, since, in a remarkable stretch starting in 1793, he patented more than eighteen separate machines, including a fire engine; the first hydraulic press, which applied the principle of feedback in the form of a self-tightening collar to prevent fluid loss; the first beer tap (apparently designed to save publicans from carrying barrels up and down stairs); a wood-planing machine that used twenty-eight tools mounted on a vertical shaft, one of which was still in use thirty years after Bramah's patent expired; and a machine for numbering and dating banknotes (the Bank of England would order three dozen of the machines).

Woodworking and fine cabinetry, however, were Bramah's original sources of income, and apparently his primary source of insecurity, since he spent at least three years working on a new method of locking his workshop for the night. He received his first lock patent in 1784, for "a LOCK, constructed on a *new* and *infallible* Principle, which, possessing all the Properties essential to Security, will prevent the most ruinous Consequences of HOUSE ROBBERIES, and be a certain Protection."

The lock made Bramah's fortune. London's upper classes were, with a good deal of justification, fearful of theft, and thus prepared to pay the equivalent of a laborer's annual salary for an unpickable lock. His marketing brilliance and intuitive understanding of branding meant that people also valued the status conferred by ownership; several decades later Charles Dickens made a point of reminding his readers that the Gray's Inn offices of Mr. Perker, the "cautious little" attorney in *The Pickwick Papers,* were secured by a Bramah lock.

However, the secret to its obduracy was a highly complicated design requiring up to a hundred separate metal pieces, each of which needed to be made to extremely accurate tolerances, and producing even the simplest Bramah locks in anything approaching the quantity demanded presented huge technical difficulties. This was a central challenge of this period of frantic invention: how to produce complicated machines in quantity.

The combination was the problem. Navigational instruments, and especially clocks and watches, were already made not only precisely, but identically; however, their market was small enough to be met on a bespoke basis. "Machines" such as buckles and millstones could be produced in bulk, but had few moving parts and a lot of room for error.

Two developments were needed to turn the custom-made machine into one that could be manufactured in quantity. Darby, Huntsman, Cort, and their competitors had tackled the first part of the problem by producing a regular supply of iron, an affordable material that could be fabricated into as many shapes as wood, but unlike wood would keep its shape even when abraded or stressed. The second was some way to turn that iron into shapes *precisely* and *consistently.* If the most important invention of the Industrial Revolution was invention itself, the automation of precision has to be one of the top three.

Appropriately enough, the development can be dated with a reasonable amount of accuracy, to the day in 1789 that the forty-year-old Bramah first met an eighteen-year-old metalworking prodigy from the Royal Arsenal at Woolwich named Henry Maudslay.

MAUDSLAY WAS BORN AT Woolwich, where his father was employed as a laborer making munitions for the Royal Navy, and soon enough the Arsenal offered employment to Henry as well, who became a "powder monkey" loading gunpowder into shells at the age of twelve. From subsequent events, we can safely assume that he was not only clever with his hands—on the sly, he forged highly prized kitchen implements for families throughout Woolwich—but careful as well, since he was in full possession of all his fingers when he left Woolwich for London six years later.

The lock design was then five years old, but the need for handcrafting made it more of a curiosity than a commercial success. Bramah was well aware of the problem, and he sought a solution from London's best-regarded blacksmith, a German named William Moodie. Moodie was stumped, but he had heard some of the legends that were already forming around Maudslay at Woolwich and recommended him to Bramah, who satisfied himself that the onetime powder monkey was not only resourceful, but even more fanatical about precision than he was himself.

Satisfying that mania for precision wasn't particularly easy in the last decades of the eighteenth century. Despite the enormous number of innovations that had appeared over the preceding eighty years, at the moment when Bramah met Maudslay, the most critical precision instrument in the heavy metal trades remained a good file; a diligent metalworker was still measured by his ability to take the rough edges off everything from a pocket watch's winding screw to the turnscrew on a one-ton cannon.

The screw is one of the canonical simple machines, along with the lever, wedge, and inclined plane, with a history dating back to the Fertile Crescent, and throughout antiquity wood screws were used as presses for oil, wine, and (by the time of the Inquisition) the occasional human thumb. Yet Maudslay sensed that the screw, despite its antiquity and simplicity, was also the tool of the future. For centuries, screws had been produced using the hand tools known as taps and dies: the former to cut interior threads—the "female" side—and the latter the exterior, or "male," threads. But until the fifteenth century, almost all of them were cut out of wood, which didn't do much for either their precision or their durability. Even when European metalworkers started to make metal screws, their dependence on hand tools required that the metal be relatively soft—and, more important, kept the screws rare. Until screws, and other sym-metrical metal objects, could be made by a machine, they were destined to stay that way.

The machine that finally broke the logjam—the lathe—was nearly as old as the screw. It had evolved considerably during the preceding two millennia, from the Egyptian "rope bow" operated by a person pulling back and forth on a rope attached to the workpiece, to the Roman bow lathe, and finally to the spring-operated pole lathe, which represented the state of the art in wood turning from about the first century CE through the Middle Ages.

Wood turning has its own satisfactions. As anyone who has ever taken what used to be called shop class will recall, there is a beauty in watching a lathe (or its close cousin, a potter's wheel) at work. The transformation of an irregular shape into a symmetrical one seems to feed what may be a universal love of harmony, and even the simplest lathe work—turning a block of wood into a chair leg or a baseball bat—is satisfying to watch, and even more satisfying to do. But it is limiting. The lathe remained a tool for creating only beauty until the sixteenth century, when it finally became a tool for creating other tools—and, most particularly, a tool for creating metal screws.

That was when the onetime wood turning machines acquired the mandrel—a spindle onto which the workpiece was attached, thereby transmitting rotation to the spindle rather than to the piece itself. Even more important, the sixteenth-century lathes finally started using the long leadscrew, which moved the workpiece horizontally as it was rotated. The leadscrew, combined with a steady platform for the cutting tool, was now the measure of precision in lathe work: Any piece turned on a lathe that used a leadscrew could be made as precisely as the leadscrew itself.

Using a leadscrew made any lathe work more precise, but its really revolutionary application emerged when a number of innovators figured out how to angle the cutting head to incise a continuous helical groove onto a smooth cylinder: to machine a screw. And not just a screw fastener; the reason lathes are frequently called history's "first self-replicating machines" is that, beginning in the sixteenth century, they were used to produce their own leadscrews. A dozen inventors from all over Europe, including the Huguenots Jacques Besson and Salomon de Caus, the Italian clockmaker Torriano de Cremona, the German military engineer Konrad Keyser, and the Swede Christopher Polhem, mastered the iterative process by which a lathe could use one leadscrew to cut another, over and over again, each time achieving a higher order of accuracy. By connecting the lathe spindle and carriage to the leadscrew, the workpiece could be moved a set distance for every revolution of the spindle; if the workpiece revolved eight times while the cutting tool was moved a single inch, then eight spiral grooves would be cut on the metal for every inch: eight turns per inch.

Thus a leadscrew that was accurate to within ½″ could operate a lathe that could cut a new screw accurate to within 7/16″, and the new leadscrew could in turn produce one accurate to within 3/8″, eventually achieving a very high degree of precision.

Not, however, high enough for Bramah's locks. For while the leadscrews on those lathes were made of iron, most of the lathe's other components were made of wood. And wood, even very hard wood, shakes. It shakes enough, in fact, that even the sharpest steel blade couldn't make a cut accurate to within 1/16″—an enormous error margin in a three-inch lock.

Henry Maudslay's first, and probably greatest, contribution to the Bramah Lock Co., and to the Industrial Revolution, was his realization of the huge advantages of a lathe made entirely of iron. Not just the leadscrew and the slide rest—a platform that holds the tool post and moves the tool laterally as precisely as the leadscrew moved the workpiece horizontally—but all the platforms, bits, and supports of the lathe. Maudslay's design integrated all of them in a manner that achieved a degree of precision greater than any could offer individually. The advantage of iron over wood turned out to be critical.

Maudslay's perception about the superiority of iron may have come to him in the form of a Glasgow Green sort of insight, but he left no diary that would confirm its origin, or even precisely when he produced the first all-iron lathe for Bramah, though it was certainly in operation by 1791. During the 1790s, Maudslay's key insight—that stability equaled precision, and iron was stable—was incorporated into a number of other tools he built for Bramah's lock business, including drills, planing machines, and possibly even a rotary file: essentially one of the first mechanical milling machines, used to shape metal into nonsymmetrical shapes, just as lathes formed them into symmetrical ones. In addition, he is credited with inventing a self-tightening leather collar that made Bramah's hydraulic press a working proposition.

However, Bramah was in the business of selling locks, not lathes, and he determined that the best business decision was to patent the things made by his new machine tools, not the tools themselves, which he kept secret as long as possible; as a result, it is rather difficult to document when, precisely, Maudslay and Bramah put them on the company's production line. No such problem exists in documenting Maudslay's devotion to his employer, which was far greater than his employer had for him. Even when Bramah promoted Maudslay to shop superintendent in 1798, he was still paying him a fairly modest thirty shillings a week, which was not enough for Maudslay's growing family. Unable to persuade Bramah to part with a living wage, Maudslay left and set up shop on Oxford Street in London, where he was employing eighty men by 1800, and nearly two hundred by 1810.

And still he remained obsessed with screws and screw-making machinery.* Maudslay rightly realized that the match of screw

* In Yorkshire, extraordinarily precise lathe work was being performed by the scientific instrument maker Jesse Ramsden, who was able to cut a screw with an awe-inspiring 125 turns per inch—that is, a Ramsden screw rotated 125 times before it traveled a single inch, which allowed for *very* fine adjustments—but his achievements in telescopes and quadrants, like those of clockmakers, though known to Maudslay, were peripheral to industrialization.

lands to bolts or receivers was key to fastening metal pieces in large machines as well as small, and that the large machines represented a far more profitable market. He spent the decade after leaving Bramah building lathes that could produce screws of any desired pitch using the same leadscrew, and that were both large enough to make the linkage for a 48-inch cylinder steam engine and precise enough to make the quarter-inch valves that controlled their operation.

The key was the leadscrew, which reproduced its exact pitch on the material to be threaded—and introduced exactly the same inaccuracies. If the leadscrew was accurate to (for example) 1/4″, then the screw, or screw fitting, it cut might make eight turns in anything from 7/8″ to 1 1/8″; if it was accurate to 1/16″, then its fittings would make the same eight turns somewhere between 31/32″ and 1 1/32″. Ten years of experimentation with different combinations of gears and cutting tools eventually resulted in a seven-foot-long brass leadscrew that was accurate to within less than 1/16″.

By then, however, 1/16″ might as well have been a foot; in another example of one problem's solution creating a new set of problems, the more accuracy Maudslay had, the more he needed. This was because, while he was iterating his way to his prize leadscrew, he had also built himself a tool that was to eighteenth-century metalworking what Galileo's telescope was to fifteenth-century astronomy. Perhaps unsurprisingly, the tool was used not to make things, but to measure them.

Micrometers, devices for measuring very small increments, were then only about thirty years old; James Watt himself had produced what was probably the world's first in 1776, a horizontal scale marked with fine gradations and topped with two jaws, one fixed and the other moved horizontally by turning a screw. With a pointer on the movable jaw, objects could be measured extremely accurately, up to 1/100″. But Maudslay's micrometer, which he nicknamed "the Lord Chancellor," was capable of measuring differences of less than 1/1000″ (some say 1/10,000″). When he measured the seven-foot-long brass screw, inch by inch, with the Lord Chancellor, he found that his "perfect" screw was actually inconsistent along its entire length: one inch might have fifty threads, another fifty-one, a third forty-nine, and the only reason it seemed accurate was that the irregularities had canceled one another out. This was clearly unsatisfactory to a perfectionist of Maudslay's degree, and the screw was recut, again and again, until even the Lord Chancellor could find no error.

There is a mythic quality to the Lord Chancellor—an Excalibur of measurement, slaying the dragon of imprecision—that explains its ubiquity in stories about Maudslay and his entire era. But that very quality tends to hide its real importance. The Industrial Revolution, however it is defined, depended on Maudslay's micrometer, and instruments like it, just as much as it did on laws protecting intellectual property or the birth of scientific experimentation. This is because *sustained* innovation is *incremental* innovation, and those increments are usually very small: a valve that weighs a fraction of an ounce less, a linkage that reduces coal consumption by a few pounds a day. Without instruments that could measure such small improvements in performance, invention was doomed to be rare and erratic; the mania for precision that was Maudslay's defining characteristic made it commonplace.

Maudslay's own inventions are impressive enough. In 1805, he patented a machine that could print designs on cotton; in 1806, he invented a new method for lifting weights with a differential motion; and in 1807, he devised a new and compact framework for supporting the cylinder of a steam engine, which permitted the use of so-called "table engines" in far smaller factory areas.

His influence is, however, larger than that, beginning with the astonishing number of other equally obsessive engineer-inventors to whom he was teacher and mentor. The best known may be the almost embarrassingly prolific Richard Roberts, who acquired patents the way Balzac wrote novels.* Another of Maudslay's as-

* Compulsively, and often brilliantly. A single one of Roberts's cotton spinning machines warranted no fewer than eighteen separate patents. His plate punching machine, operated by the same Jacquard system originally used for weaving, was the first digitally operated machine tool.

sistants, Joseph Whitworth, developed a measuring system accurate to one-millionth of an inch. This is not a misprint; until the United Kingdom joined the metric system, the standard unit for screw threads was the BSW, which stands for British Standard Whitworth.

Whitworth worked for Maudslay at the same time as Joseph Clement, who would later build, on the instructions of Charles Babbage, the prototype for the original "difference engine" —the world's first mechanical computer. Maudslay's son, Joseph, later became a brilliant marine engineer, patenting the double-cylinder marine engine that was widely used during the nineteenth century. Yet another Maudslay graduate, James Nasmyth, the inventor of a steam hammer that made him the wealthiest of them all, wrote floridly of his mentor, "the indefatigable care which he took in inculcating and diffusing among his workmen, and mechanical men generally, sound ideas of practical knowledge, and refined views of construction, has rendered, and ever will continue to render, his name identified with all that is noble in the ambition of a lover of mechanical perfection." More prosaically, though probably just as accurately, one of Maudslay's workmen remembered that "it was a pleasure to see him handle a tool of any kind, but he was quite splendid with an 18 inch file."

However, even if Maudslay had never built an iron lathe to make Bramah his locks, never become the era's icon of precision, or never turned his Oxford Street workshop into the place where, as one modern historian put it, "a 'critical mass' of inventive activity" was achieved, he would still have earned a large place in industrial history. And he would have earned it by building blocks.

A PILGRIMAGE TO BRITAIN'S sacred sites of industrialism would certainly include Boulton & Watt's Soho Foundry and the Ironbridge Gorge Museums; probably the four-hundred-foot-deep mine at the National Coal Mining Museum in Yorkshire; and possibly the pumping station at Crofton. It would certainly be incomplete without a visit to the eastern shore of Portsmouth Harbor, where the Royal Navy's largest dockyard houses museums, historically important ships, and the Portsmouth Block Mills,* the place where Henry Maudslay's machines would make up the world's first true steam-powered factory.

The quadrupling of the Royal Navy during the eighteenth century, like the Apollo space program of the 1960s, created a massive customer for technological innovation. This was, after all, the Age of Sail, and while guns and ammunition might be metal, getting those guns to where they could be useful required the pressure of wind on canvas. Wind was a useful source of power for mills, but its directional variability made it a capricious sort of transportation "fuel," and a staggering amount of human ingenuity was required to make the wind blowing *this* way drive a three-thousand-ton ship *that* way. Each sail—a three-masted ship had at least nine—was raised, lowered, reefed, and turned by some portion of more than twenty miles of rope, every foot of which ran through up to a dozen different pulleys, contained within blocks of wood. The blocks consisted of shells, usually of elm, cut with several oblong slots, or mortises, each containing a hardwood pulley fitted with metal bushings spinning around a pin, usually made of iron. A single ship of the line required as many as fifteen hundred such blocks, ranging in length from three inches to three feet, and with nearly a thousand ships at sea by the beginning of the nineteenth century, wearing out their blocks at a rapid rate, anyone who could produce them in quantity was going to make an indecent amount of money doing so; in 1800, the navy was paying more than 8 guineas—two months of a skilled laborer's salary—for a single 38-incher.

For centuries, blocks—shell, pulley, and pin—had been made by hand, with all the cost in time and error that implied. The first block makers to mechanize were the Walter Taylors (father and

* Or it would be if the Mills wasn't actually still part of the naval base itself and consequently off-limits to most visitors.

son) in 1754; eight years later, they patented their "Set of engines, tools, instruments, and other apparatus for the Making of Blocks, Sheavers, and Pins." In 1786 they received another, for lubricating the apparatus. More significant was Samuel Bentham, whose 1793 patent for woodworking machinery included a rotary planer; a circular saw; a primitive router for dovetail grooving; bevel saws, crown saws, and radial saws; radial and reciprocating mortise machines; guides, grinders, gauges, and tables; and his own version of the slide rest lathe. The Bentham application was so frighteningly comprehensive, covering all conceivable aspects of mechanized woodworking, that the Patent Office regarded it as "a perfect treatise on the subject."

An innovative naval officer determined to reform his tradition-minded service, Bentham had become fascinated by the potential for machine production of naval components largely by accident. In 1786, while in Russia, where he had gone to take a job as a naval engineer, he was so short of skilled craftsmen that the only way to produce blocks, tackles, belaying pins, and all the other wooden impedimenta of the Age of Sail was to make the process simple enough that they could be manufactured by even illiterate and untrained serfs. Or, even better, by machines.

Bentham's older brother, the political philosopher Jeremy, had independently developed an interest in woodworking by unskilled laborers. While he is best remembered for his utilitarian philosophy—"the greatest good for the greatest number"—Jeremy Bentham probably spent as much time thinking about prison reform as anything else,* and his fascination with prisons extended to the idea that woodworking was the perfect way to occupy the idle but untrained hands of prisoners.

In 1795, the Bentham brothers put their ideas together and

* Jeremy Bentham's "Panopticon," a multilevel prison with a central core from which guards could watch every move each prisoner made, and which has become a metaphor for the modern surveillance state, is one of his best-remembered, if creepiest, ideas. Less well known is that the original Panopticon was designed by Samuel Bentham, for use in supervising laborers at Krichev, the estate of Prince Vasiliy Potemkin.

drafted a contract proposing that the Admiralty use prison labor to operate the woodworking machines used to produce naval stores. Jeremy, evidently ambitious to find an even larger market for Samuel's inventions, wrote a letter to his friend, the Duc de la Rochefoucauld, a French nobleman then in exile in North America, asking whether "*a Propos* of my brother's inventions, do you know of anybody where you are . . . who would like to be taught how to stock all North America with all sorts of woodwork . . . on the terms of allowing the inventor [i.e. Samuel] a share of the profits as they arise?" The Frenchman evidently had no immediate help on offer, but a few years later, the letter was apparently the chief subject of discussion at a dinner party in New York City, whose guests included the recently resigned secretary of the Treasury, Alexander Hamilton, and another French émigré, a former sailor and engineer named Marc Isambard Brunel.

BRUNEL HAD BY THEN crowded a fair bit into his first twenty-seven years. When he was eleven, he traveled from his birthplace at Haqueville in Normandy to attend the seminary of St. Nazaire in Rouen, where soon enough the priests realized that the boy's vocation was more mechanical than pastoral. They sent him to live with the American consul in Rouen, a retired sea captain, to be educated in hydrography and drafting as preparation for a naval career. He was commissioned into the French Navy in 1786 and served on a dozen voyages, but when he returned from the West Indies in 1792, France's three-year-old revolution was taking a violent turn. In the fall, Parisians had imprisoned the king and queen, and the era of mass executions known as the Reign of Terror was looming. Brunel decided to emigrate. His education seems to have nurtured interests in navigation and the United States of America equally; a year later, in September 1793, he landed in New York.

The royalist-leaning Frenchman was enthusiastically welcomed by the new republic. On the recommendation of two of his fellow passengers, he almost immediately won a job to survey a land grant near Lake Ontario and drew up plans for a canal between Lake Champlain and the Hudson River. Once he took on U.S. citizenship, he was named chief engineer for the City of New York, building foundries, laying out roads, and planning for the defense of the city's harbor, which was the job he held when the subject of block making was broached at the dinner with Hamilton.

In Brunel's later recollection, the flash of insight that followed struck him as he was "roaming on the esplanade of Fort Montgomery." Just as with James Watt's stroll on Glasgow Green, a machine had appeared to him, more or less fully formed, in which the mortises in the blocks could be cut by chisels moving up and down in series "two or three at a time" as the blocks were conveyed along a moving line.

On January 20, 1799, Brunel sailed for Britain armed with a letter of introduction from Hamilton to the First Lord of the Admiralty* and a patent specification for his machine. The First Lord arranged an immediate introduction to Samuel Bentham, by now Inspector General for Naval Works, but it took two years before he finally received a patent for his "New and Useful Machine for Cutting One or More Mortices Forming the Sides of and Cutting the Pin-Hole of the Shells of Blocks, and for Turning and Boring the Shivers."

Brunel had broken block making into a series of steps that synchronized a dozen different woodworking processes. During one of those steps, machine-driven chisels—between one and four—cut out slots in a rectangular block of wood, while their reciprocating motion drove a gear that moved the block laterally the length of the needed mortise. In another, sheaves—the pulleys intended to fit inside the mortises—were made by a rounding saw that made a circular disc while simultaneously cutting a groove in the middle. Bentham's own designs were ingenious enough, but the machines specified in his 1793 patent operated indepen-

* Earl Spencer of Althorp, a direct ancestor of Diana, Princess of Wales.

dently; each one could typically complete only a single step. Brunel's plan, in essence, took the motion imparted by one machine and used it to drive another. By requiring that each step in any procedure be driven by the preceding one, he effectively automated the entire block-making system.

In the same year that the patent was awarded, Brunel persuaded Samuel Bentham to put his ideas to work at the navy's largest dockyard, in Portsmouth. For that, he needed a toolmaker. Someone like, for example, Henry Maudslay, whom Bentham hired in 1802 to turn Brunel's drawings into machines.

Or, more accurately, to revise them. The brilliance of Brunel's patented idea was the manner in which it coordinated the different cutting and drilling movements, but their very coordination demanded precision that could be measured in thousands of an inch. The design, however, specified the use of wood for dozens of components, and vibration alone introduced errors larger than that. Maudslay, who by then knew more than anyone else living about eliminating vibration, did so by translating Brunel's designs into cold iron. His machines—he ended up building forty-five or so—included power saws for roughing out pieces of elm into useful sizes; drills, mortising chisels, and scorers; rotary saws for the sheaves; and even forging machines to make the iron pins and bushings. And they were all, in the end, made of the same cast iron that had become so reliably available.

Maudslay's fee for constructing the machines came to the very handsome sum of £12,000—considerably more than $1 million in current dollars—which made sense only given the contract that Brunel had executed with Bentham and the Admiralty. That agreement guaranteed that the machinery would allow six men to do the work of sixty, with annual cost savings in the neighborhood of £24,000, which would be used to calculate his own payment. The final accounting is almost incomprehensible—in 1808, the Mills produced 130,000 blocks at a nominal price of £54,000, but the figures that were used to calculate the annual "royalties" payable to Brunel came out anywhere between £6,691 and £26,000—and, perhaps predictably, Brunel had to chase down the money he was owed. He didn't come close to recouping his own investment until 1810, when the Admiralty settled on a single payment of a bit more than £17,000.

By then, the Portsmouth Block Mills had become Britain's most advanced industrial factory and among its most important defense plants. On the day in 1805 that Horatio Nelson left Portsmouth in search of the French fleet he would eventually find at Trafalgar, his last stop was the Mills. Three years after Nelson's death, the Portsmouth Block Mills was producing "an output greater than that previously supplied by the six largest dockyards." Just as important, it had become Britain's best advertisement for the virtues of industrialization. The Portsmouth Block Mills was extraordinarily public, and deliberately so. Bentham had hoped to publicize the machine age by making the mills open to the public (a cause for much complaint by the engineers), and he encouraged articles about it in numerous journals and encyclopedias, including six consecutive editions of the *Encyclopaedia Britannica.* To the degree that the machines of the Industrial Revolution depended upon awareness of, and inspiration from, other machines, Henry Maudslay's saws, drills, and chisels earned Portsmouth Block Mills its place on the pilgrimage route.

So did the engines that drove them.

Even before Bentham had put Brunel's ideas and Maudslay's hands to work, he had shown a powerful affection for novelty in both naval tactics—he is justly famous as an early advocate for replacing solid shot with explosive shells in naval combat—and engineering. In 1798, he introduced at Portsmouth the Royal Navy's first stationary steam engine, a relatively small "table engine" built and installed by James Sadler, a member of Bentham's staff, to drive one of the early rotary saws. That one was supplemented in 1800 by a Boulton & Watt beam engine housed in a separate building, despite the Navy Board's nervous belief that these newfangled machines would "set fire to the dockyards [and] would occasion risings of artificers, and so forth."

Though the first engine to drive Maudslay's saws and chisels came from the Soho Foundry, it wasn't to be the last. The year it was installed, 1800, was also the year that the patent for the separate condenser finally expired, thus in theory opening the marketplace to competition; and indeed, in 1807, the Royal Navy replaced its Boulton & Watt machine with a more powerful substitute from the company's most serious challenger, Matthew Murray.

MURRAY WAS THEN ABOUT forty years old, like so many others a product of the apprentice system—in his case, to a "whitesmith" or tinker in his home of Newcastle-on-Tyne—who become a journeyman mechanic and inventor, first in the employ of a linen manufacturer named John Marshall, then in partnership with two friends named James Fenton and David Wood. In 1797, the new company, Fenton, Murray, and Wood, patented a brilliant new steam engine design, one that used a horizontal cylinder and incorporated a new valve, designed by Murray, that dramatically improved engine efficiency.

By this time, an awful lot of the big stuff in steam engine design—the separate condenser, the double-acting engine—had already been introduced, patented, and seen those patents expire. Each time this happened, the innovation in question lost its competitive advantages, with the result that the search for smaller and smaller improvements was well under way. Even so, some small improvements resulted in large profits, and one was certainly Murray's D-valve (so called for its shape), which controlled the flow of steam. Earlier self-acting valves had been relatively heavy and required a not inconsiderable amount of the engine's own steam power to lift—and every bit of energy that went into lifting a valve was not available for any other work. The lighter the valve, the more efficient the engine, and the D-valve weighed less than half as much as its predecessor. The valve's shape was likewise a cost saver: it absorbed less heat than its predecessor, thus increasing engine efficiency, since every bit of heat used to heat the engine parts was no longer available to make steam.

There was no doubt of the originality of the D-valve, but in 1802, Murray patented "new combined steam engines for producing a circular power . . . for spinning cotton, flax, tow and wool, or for any purpose requiring circular power," and this one was challenged in court by Boulton & Watt. Their victory (on a technicality: Murray had included dozens of improvements in the same patent application, and the law provided that if any one of them was not completely original, it invalidated the entire application) did not endear them to Murray. After his loss, he spent large sums advertising his originality, attempting to persuade Britons that his ideas weren't stolen from Watt. The experience enriched the newspapers but soured him on the patent system, which he rarely used again. Even so, the conflict continued; he and his partners planned to expand their factory in Leeds, only to find out that Murray's conflict with his Birmingham competitors did not improve with time; the planned expansion of Fenton, Murray, and Wood's four-story-high circular factory—the famous "Round Foundry" in Leeds—was thwarted when Boulton & Watt bought up every surrounding acre. It soon became clear that the cultural and legal revolution that had transformed ideas into an entirely new sort of valuable commodity had created a new sort of conflict as well. The availability of patent protection was, predictably, motivating inventors to make more inventions; it was also motivating them to frustrate competing inventions from anyone else.

The argument between those who believe legal protection for inventions promotes innovation or retards it continues to this day. For both sides of the debate, Exhibit A is often the litigation between James Watt and Jonathan Hornblower.

HORNBLOWER, THE SON OF a onetime steam engine mechanic (the steam engines in question were reputedly Newcomen's) and nephew of another,* followed them into the family business

* Uncle Josiah emigrated to America in 1753, where he became a judge and speaker of the House of Assembly in New Jersey before dying in Belleville, New Jersey, in 1809.

when he hired on with Boulton & Watt to install engines in Cornwall in the late 1770s. By 1781, either by native ingenuity or careful observation, he was able to draft a patent for a revolutionary new kind of steam engine that coordinated two separate cylinders, one at higher pressure than the other, and used the pressure exhausted from one cylinder to drive the other. This both increased the machine's output by as much as a third, and, by running each cylinder in a sort of syncopated rhythm, reduced the "dead spots" where the piston reversed direction (this is known as "smoothing out the power curve"). In addition, Hornblower's "compound engine" also incorporated a couple of less revolutionary items: a separate condenser and air pump, both of which were still protected by Watt's original patent, to say nothing of the 1775 extension.

The new design did not catch on immediately. The patent itself was vague enough that most of what was known about the Hornblower engine was little more than speculation. It wasn't until 1788 that Watt caught wind of a speech given by Hornblower to a group of Cornish miners, in which his onetime employee reputedly said that Watt had not even invented the separate condenser, and that they were in consequence paying Boulton & Watt an unnecessary royalty. The speech offended Watt, but it did not, by itself, threaten the dominant position of his engine design. Three years later, however, Boulton & Watt were growing concerned about both potential competition and actual infringement. In November of 1791 Watt wrote to Boulton that "the ungrateful, idle, insolent Hornblowers [there were three Hornblower brothers, none of whom had particularly endeared himself to Watt] have laboured to evade our Act, and for that purpose have long been possessed of a copy of our specification." In 1796, they sued Hornblower and his partner, David Maberley, asserting infringement on the separate condenser patents.

The most illuminating aspect of the entire affair was the difference in the way that Watt and Boulton viewed it. For Watt, the theft (as he saw it) of his work was a deeply personal violation. In

1790, just before realizing the extent of what he perceived as Hornblower's theft of his own work, he wrote,

> if patentees are to be regarded by the public, as . . . monopolists, and their patents considered as nuisances & encroachments on the natural liberties of his Majesty's other subjects, wou'd it not be just to make a law at once, taking away the power of granting patents for new inventions & by cutting off the hopes of ingenious men oblige them either to go on in the way of their fathers & not spend their time which would be devoted to the encrease [*sic*] of their own fortunes in making improvements for an ungrateful public, or else to emigrate to some other Country that will afford to their inventions the protections they may merit?

Despite his own confidence, he was aware of the complicated public relations aspect of his situation: "Our cause is good," he wrote, "and yet it has a bad aspect. We are called monopolists, and exactors of money from the people for nothing. Would to God the money and price of the time the engine has cost us were in our pockets again, and the devil might then have the draining of their mines in place of me. . . . The law must decide *whether we have property in this affair or not*" (emphasis added).

In the event, the law did decide against Hornblower, and in favor of Boulton & Watt (though not until January 1799, only a year before the expiration of Watt's patent) notwithstanding the testimony of none other than Joseph Bramah, the lockmaker, who stated under oath that Watt's separate condenser offered no improvement on Newcomen's engine, referring to the 1769 innovation as "monstrous stupidity," which rubbed Watt very raw indeed. Boulton, who had only money at risk rather than pride, recommended keeping an even keel: "I think we should confine our contentions to the recovery of our debts, and in that be just, moderate and honourable, for sweet is the bread of contentment."

The lessons to be derived from the Hornblower litigation are probably fewer than generally thought. The lawsuit has been used to underline the many contemporaneous perspectives on intellectual property; the abusive character of patent law; and even the geopolitics of eighteenth-century Cornwall. It certainly is not an object lesson in the wages of patent theft; despite numerous citations that find him ruined and even jailed for his troubles, Jonathan Hornblower actually ended up quite wealthy, and continued to pursue patents for steam engine improvements on his own for years.

If the case informs anything, it is actually the great virtue of an environment that recognized the value of intellectual property. Whatever the failures of any specific judicial remedies, a society that wants good ideas to triumph over bad—for superior technology to replace inferior—must promote the creation of as many ideas as possible. In the end, the compound engine was fairly rapidly "rediscovered" by a onetime carpenter named Arthur Woolf, who patented, in 1804, a new method of using steam in an expansive engine, this time by raising the temperature of the steam within the cylinder instead of the boiler, thus creating "a sufficient action against the piston of a steam engine to cause the same to rise in the old [Newcomen] engine . . . or to be carried into the vacuous parts of the cylinder in [Watt's] improved engines." That is, it discharged steam directly from a higher-pressure cylinder into a lower-pressure one, thereby compounding the power stroke. With the separate condenser now in the public domain, he was free of the risk of litigation.

By 1808, a compound engine had made its way to the Portsmouth Block Mills, though both Brunel and Maudslay had moved on to other ventures. The former ultimately went bankrupt, despite being paid more than £17,000 by the Navy, and in 1821 was incarcerated as a debtor. He had once again gone to the patent well, in August 1810, with a machine for mass-producing shoes and boots for the army, and when peace came after Waterloo in 1815, he was left with truckloads of unsold footwear, a reminder of the fickle nature of a fortune built on government contracts. His most ambitious endeavor, a tunnel under the Thames, would ultimately be completed by his even more famous son, Isambard Kingdom Brunel.* Maudslay, on the other hand, would eventually build engines for forty-five naval ships, including HMS *Lightning,* the Royal Navy's first steamship, and HMS *Enterprise,* the first to steam to India.

Maudslay's lasting fame, however, came from his work on his beloved lathes, where he applied the precision of one-at-a-time scientific instrument making to engineering for mass production—or what passed for mass production in the eighteenth century. While the world of British invention was forming up on the pro- versus anti-Watt debate, Maudslay stayed above it, respected, even beloved, by everyone. His 1831 epitaph read, "A zealous promoter of the arts and sciences, eminently distinguished, as an engineer for mathematical accuracy and beauty of construction, as a man for industry and perseverance, and as a friend for a kind and benevolent heart."

By then, the biggest transformation of all was well under way in Britain's steam-driven economy: In the first decades of the nineteenth century, the factories that Boulton had promised to power with Watt's engines were manufacturing not iron, or wood, but cloth.

* Isambard Kingdom Brunel is so famous, in fact, that a 2002 BBC poll to select the one hundred greatest Britons placed him second, behind Winston Churchill, but ahead of Shakespeare, Darwin, and Newton (to be fair, so was Princess Diana; Watt came in eighty-fourth, behind such immortals as Michael Crawford, David Beckham, and Boy George). One result is that his father, who preferred being called Isambard during his own lifetime, is now known as Marc.

CHAPTER TEN

TO GIVE ENGLAND THE POWER OF COTTON

concerning the secret of silk spinning; two men named Kay; a child called Jenny; the breaking of frames; the great Cotton War between Calcutta and Lancashire; and the violent resentments of stocking knitters

THE CITY OF LIVORNO sits at the northern end of Italy's resort-spotted Etruscan coast, overlooking the Ligurian Sea. At its center is the original town: a walled compound, made up of two separate forts that in the year 1715 were enough for Livorno to serve as one of the most important ports in the Mediterranean. Livorno's walls, its fortifications, even its streets and canals, were a sixteenth-century bequest from the city's Florentine rulers, the Medici family, who had also bequeathed to the city its cosmopolitan air and legendary hospitality to foreigners. The city's constitution of 1606 granted privileges and immunities to immigrants, including Jews, Greeks, Armenians, Dutch, and Muslims; it even attracted a large number of traders and artisans from England, who, in the distinctive manner of English away from home, renamed it Leghorn.*

* The name, which appears in even some modern maps, is not such a leap as it seems. The city was also known, in the sixteenth and seventeenth centuries, as

The name stuck, in Anglophone countries, for centuries—a style of hat and breed of chicken still carry the name—because of Livorno's large and well-known expatriate community. The profile of that community was probably never higher than in 1822, when both Byron and Shelley were residents, but a hundred years before, another Briton had made himself at home in the city, with less publicity, but more significance. John Lombe was his name; and what brought him to Livorno was silk.

LOMBE WAS THE SON of a woolen weaver from Norwich, and a onetime apprentice to another weaver in Derbyshire, where in 1702 he took employment as a mechanic for a small silk mill owned by a lawyer named Thomas Cotchett. England by then had tried and failed half a dozen times to start up its own silk industry, less from any deficiency in raw materials than from lack of the technology required to make its production economical. But neither Cotchett nor Lombe was prepared to give up on the profits to be made from selling silk cloth and garments—profits far greater than for any other fabric.

By the eighteenth century, silk had been commanding high prices for millennia; during the eighth century BCE, silk was one of the Zhou Dynasty's most widely cultivated "crops," with tens of thousands of farmers feeding white mulberry leaves to domesticated silkworms with the Linnaean moniker *Bombyx mori* before the chrysalis stage, then steaming or boiling the cocoons and pulling, or "reeling," the filaments that emerged into strands of silk. Those strands were not only very long but triangular in cross-section, which gave silk thread its distinctive reflective quality; the combination of length and lustrousness made the stuff easy to weave, and even easier to sell.

The reason this is a matter of note in the history of industrial-

---

Legorno, harking back to a pre-Roman people known as the Ligurians, either directly or because of proximity to the Ligurian Sea. The two names, despite the number of shared letters, have no etymological connection.

ization, however, has less to do with the beauty of the fiber—a fiber is, technically, anything with a length at least one hundred times its diameter—than with its structure. Virtually all textiles, from linen to rayon, are made using the same step-by-step process. First, foreign materials, if any, must be removed, or carded, from the raw fiber; then individual strands must be separated, with those of uniform length combed, or aligned in parallel. The carded and combed fiber is then twisted into yarn, or spun, and ultimately woven by interlacing yarns at right angles.

Or so the sequence goes with *staples,* such as wool, linen, or cotton, which must be drawn at an angle so that the relatively short fiber twists into a longer and stronger yarn. But reeled silk is a *filament:* a very *long* filament. A single cocoon of *B. mori* holds only a few grams of silk's two constituent proteins, fibroin and sericin, but they form threads, one-twentieth the diameter of a human hair, that can reach the length of ten football fields. The result is that silk produces yarn without either combing, or carding, or drawing; all it needs to be is tightly twisted and it's ready to serve as the warp on a silk loom.

Silk from Chinese looms started appearing in Egypt as early as 1000 BCE, but it didn't really take off as an article of trade until 50 CE, when the Han emperor made a "gift" of ten thousand rolls of the stuff to pacify the western nomads known as the Xiongnu—silk that would eventually be shipped westward across the Central Asian desert along the not yet named Silk Road to Persia and the Mediterranean empire of Rome.

China remained a major supplier of silk to Europe for centuries, but with the breakup of the Mongol empire in the fourteenth century and the rise of the Ottomans, silk production shifted west. The Turkish city of Bursa was shipping more than 100 metric tons annually by the beginning of the sixteenth century, most of it carried by Armenian merchants to either Italy or southern France. When the cities of Toulon and Marseille imposed a series of confiscatory tolls on west Asian silk in the 1650s, the free port of Livorno was happy to step into the breach and almost immediately became the entrepôt of choice for silk. In 1665, five Dutch ships left the Turkish port of Smyrna (now Izmir) for Livorno carrying five hundred bales of silk; in 1668, they carried twenty-five hundred.

The dramatic increase was driven by technology. The unique characteristics of silk fibers made them uniquely easy to weave by machine, and the demand for such machines was greatest in the triangle formed by the "silk cities" of Pisa, Lucca, and Livorno. At the very beginning of the seventeenth century, an engineer from Padua named Vittorio Zonca had designed the first machine to turn silk fiber into silk cloth. It was Zonca's machine, which consisted of two frames, one inside the other, with the outer one holding spindles and reels and the inner one rotating around a central vertical post that held the silk by friction—a machine that had been kept a secret in the Piedmont district for more than a century—that drew John Lombe to Livorno. Or, more exactly, the plans for the machine, which he had traveled from Derbyshire to acquire.

It's not known how he got them; bribery is a good guess. But when Lombe left Livorno in 1716, he had a set of plans for Zonca's mill,* and two years later, something even better. In 1718, having figuratively filed the serial numbers off his smuggled plans, he received patent number 422 for his invention of "three sorts of engines never before made or used in Great Britaine [*sic*], one to winde the finest raw silk, another to spin, and the other to twist the finest Italian raw silk into organzine in great perfection, which was never before done in this country." That's about as specific as the application got, and it's hard to avoid the suspicion that the vagueness was deliberate, along with the decision to include a lot

* In one sense, the plans were unnecessary, since the mill was described in Zonca's posthumously published 1607 book *Novo teatro di machine,* a copy of which was owned by the Bodleian Library at Oxford. Unfortunately, while there is no reason to think that Lombe, or for that matter anyone else, was aware of it, the existence of the book makes at least some of the more romantic stories about the theft slightly less persuasive. As a case in point, Lombe's first biographer, William Hutton, wrote that the Piedmontese silk weavers were so angry at the theft of their secrets that they sent a femme fatale ("an artful woman," in Hutton's words) to England to seduce and poison John Lombe.

of Italian, apparently so that the process would remain exclusively Lombe's even after the patent expired.

In 1719, John Lombe joined with his brother, Thomas, a mercer (that is, a dealer in fine, usually imported cloth) and a member of the London guild known as the Mercer's Company,* to build their own mill. The site they chose was in Derbyshire, on the same island in the River Derwent used by Cotchett, and for the same reason: the water flowing past could easily be used to power their new silk mill. The "Italian Works," as the Lombe mill was locally known, was a five-story structure, 100 feet long by 37 feet wide, set on pillars over an undershot waterwheel that drove a single vertical shaft operating machines on each of the five floors. The mill, which employed more than two hundred men, was able to produce so much silk that Thomas Lombe's investment, reported at £30,000, had increased by 1732 to more than £80,000.

Lombe neglected to point out this seemingly pertinent fact when he petitioned Parliament for an extension of his 1718 patent, arguing that "he has not hitherto received the intended benefit of the aforesaid patent, and in consideration of the extraordinary nature of [the] undertaking." Moreover, he was then locked in litigation with a group of potential competitors eager to get into the silk business, who introduced a suit, *The case of the manufacturers of woolen, linnen, mohair, and cotton yarn . . . with respect to a bill for preserving and encouraging a new invention in England by Sir Thomas Lombe.* He must have been a persuasive advocate on his own behalf, because even though the Crown declined to reward him with a patent extension, it did dismiss the suit, and paid him £14,000 in the bargain. Lombe would go on to become an alderman and a sheriff of a ward of the City of London; in 1739 he died, leaving an estate of £120,000, a poor moral lesson about the hazards of theft.

* The Mercer's Company, which dates back to at least 1348 and which was chosen by the Lord Mayor of London as the first guild in the city's hierarchy in 1515, still exists, along with one hundred other so-called "livery" companies including traditional ones like goldsmiths and weavers and rather more modern ones like information technologists.

Lombe's career is even more telling, a reminder that mechanization was a necessary but not sufficient component of national industrialization. No matter how efficient the Zonca/Lombe machine, it was still spinning yarn for a fabric whose appeal was restricted to the elite of English society. This simple fact placed the same ceiling on expansion that had limited the potential of every other innovation since Heron started making toys for Alexandria's nobles. Only one silk spinning factory—the Lombe mill at Derby—was established in England before 1750, and obviously waterpower was sufficient for all its needs. The true industrialization of Britain, and subsequently, the world, depended on a commodity that could attract consumers not by the thousands but by the millions. Something that could be produced in such quantity that hundreds of factories would need steam power not only to manufacture but (remember *Rocket*) to transport it.

Something like, for example, cotton.

TODAY, THE SEED FIBER of plants belonging to the genus *Gossypium* is the world's most important nonfood agricultural product, with something in the neighborhood of 115 million bales, or twenty-eight million tons, produced annually. All of that production comes from the plant's boll, or seed pod, which appears after the plant blossoms and, as it matures, grows "hair" in the form of fibers from two to three inches in length. Since not all bolls mature at the same time, for most of the crop's history, handpicking has proved to deliver the highest yields; this has resulted in a number of well-known consequences, including the durability of the institution of slavery, from Egypt in 3000 BCE to the American South until 1865 or so. Abusive labor practices and cotton appear together pretty much everywhere, in fact, that the climate is temperate, with a lot of moisture during the growing season and a hot and dry harvesting season. Which describes the banks of not only the Nile and the Mississippi, but the Hooghly: the river that runs past Calcutta, home to the world's first great multinational corporation.

The international venture that would ultimately be known as the Honourable East India Company was created by a royal charter—a letter patent—issued by Elizabeth I on December 31, 1600, providing a fifteen-year monopoly on trade with the so-called Spice Islands. Soon enough, the charter, and the ambitions of the Company, were to embrace south Asia and the Indian subcontinent; its scale was enough to make twenty-first-century multinationals hide their heads in shame. From about 1608 until 1757, the Company merely dominated India's economy; from 1757 until 1858, it ruled nearly half the subcontinent as sovereign, with its own tax collectors, police force, and army.

India had many attractions for the Company, but by far the largest was that India could produce more cotton more cheaply than anywhere else on the planet. Even before the Company chose the village of Calcutta on India's east coast as its trading post in 1690, they were in the cotton business; shipments of Indian "calico"—named not for Calcutta, but for the Malabar Coast entrepôt known as Calicut—rose twentyfold between 1620 and 1625 and another fivefold between 1625 and 1690.

Production increased to meet demand, and demand for Indian cotton rose because it was not only cheaper than cotton from elsewhere, but better—the result not of superior technology but of a gigantic labor pool with centuries of expertise. English cotton thread was not only pricier than Indian, but too weak to be used on its own; because weaving used the vertical threads of the warp to hold the lateral threads of the weft in a lattice, the warp fibers needed to be both longer and stronger. This obliged English weavers to use local cotton only in combination with much stronger linen, to make the cloth known as *fustian*. Even then, it made for a very rough weave indeed, not nearly smooth enough to accept the printed designs demanded by Britain's aristocrats, or the increasingly prosperous British middle class.

As Indian cotton began to crowd out not only domestic cotton but all domestic cloth, British textile manufacturers predictably sought protection from imports. They did not count any large number of cotton weavers, given the small part that the fiber was then playing in the English economy; but they did include politically powerful weavers of wool, and especially of silk. In particular, the hand silk weavers of Spitalfields, a hamlet in the East End of London, pressured Parliament to pass the first of what would be known as the Calico Acts.

The Calico Acts (the first was passed in 1700, the second, more restrictive one twenty years later) prohibited both the import and ownership of Indian printed cottons. It was a decisive victory for the large but dispersed English textile industry against the single largest joint stock corporation in the kingdom. But if the Acts were originally drafted to favor manufacturing at the expense of trade, they failed miserably. In one of history's most significant validations of the law of unintended consequences, the Acts, which originated as protection for Britain's woolen, silk, and linen manufacturers, sheltered the nascent cotton industry even better. The results were, to understate the case, startling, beginning with the creation of the most valuable export industry in human history. Between 1700 and 1750, British export trade in textiles doubled; by 1800 it had trebled, and, with two-thirds of the total generated by cotton goods, British manufactured exports amounted to 10 percent of national income—the largest percentage ever enjoyed by any nation before or since.

The explosive growth in the output of Britain's textile manufacturers was fueled by the equally explosive growth in the number of potential consumers for their products. The market for cotton, in fact, was so avid that only extraordinary increases in productivity could satisfy it. Since the domestic market could expand only as fast as the population itself, really fast growth needed to harness colonial policy to the export-friendly protectionist philosophy that would come to be called mercantilism. Thus the policy of conquering large territories was justified not because of a colony's mineral wealth, but because of its consumers.

Those overseas consumers were needed badly, because while the domestic market was growing (the British economy trebled in the century following the passage of the first Calico Act, mostly because of population growth, but partly because per capita GDP grew by a third, twice as fast as anywhere else in Europe), it wasn't growing nearly fast enough.

Not all of the increase was even measurable. Hundreds of studies of probate show dramatic increases in the inventories of furniture, clothing, household tools, and so on that Britons were bequeathing to their heirs, which strongly suggests that the stock of material goods was exploding. Even those eighteenth- and nineteenth-century British households that were seeing no increase in their cash income were nonetheless able to reallocate that income to purchase more market-supplied goods in preference to homemade. They were the ones who were able to attract the attention of a generation of inventors eager to replace their homespun with something better, or at least prettier.

And they didn't just substitute market-bought commodities for homemade; they also replaced them with products never before imagined. Throughout Britain, members of the middle and even working classes even looked different, once they were widely able to replace dyed wool with cotton prints.

In order to make cotton fabric from domestic yarn smooth enough to accept prints, however, Britain's textile manufacturers needed to master the other half of the clothmaking equation. The industry's next world-changing invention was intended not to spin fiber into yarn, but weave yarn into cloth.

With the exception of flaking stone into useful shapes (a skill that seems unlikely to return to the vocational curriculum), weaving is humanity's oldest craft. People in Mesopotamia and Turkey wove both baskets and cloth around 8000 BCE, and the first technique, simple over-and-under latticework, remained pretty much the only technique for at least four millennia. By 2000 BCE, however, far more complex weaving was being practiced, as is evidenced by models of looms found in Egyptian tombs. These frame looms had replaced the temporary "looms" made by either hanging fibers from tree branches or stretching them across a hole in the ground, making it possible to use heddles (the cables or wires used to separate the threads that form the warp of a piece of woven cloth) and permitting a shuttle to carry the weft laterally. A fabric's texture and design are created by simultaneously lifting the heddles to create a space, or shed, and interlacing the warp with different weights and colors of weft. The more heddles in a loom, the more combinations possible.

For millennia, the hands controlling those heddles and shuttles were overwhelmingly female, as, indeed, were the artisans spinning the weaver's yarn. When the fourth-century church father Saint Jerome recommended the craft for his female parishioners as an inoculation against the idleness sure to lead Eve's daughters into temptation, he was giving a Christian spin to the same occupation that Penelope used to keep her suitors at bay a thousand years before. By the sixteenth century, however, weaving had become the prototypical cottage industry: diffuse, traditional, and decidedly low-tech. It would remain so for two centuries. As with metalworking, toolmaking, and a dozen other crafts that had experienced a thousand years' worth of occasional innovation, the textile industry was able to enter into a cycle of sustainable invention only in the early eighteenth century.

THE CYCLE BEGAN WHEN John Kay, the fourteen-year-old son of a Lancashire farmer, was apprenticed to a maker of reeds (the devices used to separate the threads of a warp, usually made of cane, or reed). In legend at least, Kay left after a single month, convinced he had learned everything he would ever need to know. It seems likely that he didn't know everything, since it was eight years later, in 1726, that his first invention appeared, an improved reed that used wire instead of cane. He never patented what soon became known as "Kay's Reeds," but it was certainly not out of any antagonism to the principle of patenting. In 1730, he acquired a patent for a method of preparing the twine used for looms, and three years after that, one for a machine for dressing wool. More significantly, in the same year, Kay patented a new shuttle that was initially known as a wheel shuttle, then a spring shuttle; no one called it a flying shuttle until 1780.

Before Kay's invention, looms had been operated by weavers passing the shuttle, which carried the weft, through the warp threads by hand. As a result, any loomed cloth was going to be about the width of a human wingspan. By putting the shuttle on wheels and attaching cords to either end, Kay's invention permitted it to "fly" by pulling the cord in either direction. It would take another fifty years for its use to become widespread, but despite its relatively leisurely adoption, the flying shuttle made Kay, if not wealthy, then at least prosperous; in 1738, he described his profession as that of "inventor," but by 1745, had promoted himself to the status of "gentleman." In 1747, he moved to France, where policies toward invention and inventing were as capricious as they had been in England under the Tudors: Inventors in good odor at the Bourbon court could be rewarded with pensions, loans, production subsidies, exclusive franchises, and even titles. Kay was able to receive a French "patent" on his shuttle just as his English patent was expiring, and in 1749 a pension.

Kay may have seen himself as a gentleman, but he never stopped inventing. In 1738, he patented a windmill that successfully competed with Newcomen's atmospheric engine in raising water from mines; in 1745, a loom whose spindles were coordinated by treadles, thus permitting the weaver to keep hands free; and in 1754, a new machine for making the cards used to store weaving designs. He spent so much time, in fact, traveling from France to Britain and back again in order to defend his patents that the French government revoked his pension in 1760 (though they restored it in 1770).

Without minimizing the importance of his inventions, Kay's posthumous reputation, as changeable as a couturier's hemlines, may have even more to say about the special character of the Industrial Revolution in England. Though his death in France occurred in such vague circumstances that even the year cannot be established precisely, a hundred years later, by the 1880s, the critical importance of textile mechanization in the making of what was newly called the Industrial Revolution was so obvious that he, and other clothmaking pioneers, were being lionized in both English and French biographies. When the times, and the culture, demanded a hero, Kay, a Central Casting dream for the part of the brilliant inventor denied credit, was made to fit the bill, and his reputation has risen and fallen regularly ever since.

That is not, however, the case with the flying shuttle itself, which indisputably revolutionized the craft of weaving. It didn't do so by making the skills of the artisan redundant. Quite the opposite, in fact. Kay's flying shuttle made it possible for weavers to produce a wider product, which they called "broadloom," but doing so was demanding. Weaving requires that the weft threads be under constant tension in order to make certain that each one is precisely the same length as its predecessor; slack is the enemy of a properly woven cloth. Using a flying shuttle to carry weft threads through the warp made it possible to weave a far wider bolt of cloth, but the required momentum introduced the possibility of a rebound, and thereby a slack thread. Kay's invention still needed a skilled artisan to catch the shuttle and so avoid even the slightest bit of bounce when it was thrown across the loom.

The significance of this fact for industrialization was twofold, and instructive. The initial impact of Kay's invention was an increase in the productivity of Britain's weavers—enough of an increase that they were able to weave all the yarn that they could get in less time than ever before. And they could do it by hand. Though power looms had existed, at least in concept, for centuries (under his sketch for one, Leonardo himself wrote, "This is second only to the printing press in importance; no less useful in its practical application; a lucrative, beautiful, and subtle invention"), there was little interest in them so long as virtually all the available yarn could be turned into cloth in cottages. This fact reinforced the weaver's independence; but it also encouraged another group of innovative types who were getting ready to put spinning itself on an industrial footing.

The first tools for spinning are not quite as old as the first looms. For millennia, the device used to make yarn consisted essentially of two simple sticks; one, the distaff, held the unspun fiber, while the other, the spindle, imparted twist as a pitchfork-shaped flyer wound the fiber around its "tines" and into yarn, with all the twisting and rotating done at different, though proportionally related speeds. The first wheels used to mechanize the process by turning the spindle at a consistent rate were probably invented in India during the fifth or sixth century CE, and made their way to Europe around the end of the first millennium, though the earliest documented European spinning wheel can be dated only to 1298.

The significance of the spinning wheel is as large in the history of mechanical invention as it is in the history of textiles; it was not merely the first machine to transmit power via a belt, but after 1524, when Leonardo (surprise!) put wheel, crank, connecting rod, and treadle together, it was the first to do so with various parts of the machine revolving at different speeds. The bigger the wheel, the larger the demand for power; the wheels of the fourteenth century were spun by hand, by the sixteenth, by a foot treadle. Inevitably, more demand for yarn meant more demand for power, either animal, wind, water, or, eventually, steam.

The first step in that direction was taken in 1738 by Lewis Paul, a onetime carpenter who patented a machine that cleaned and carded fiber and "put [it] between a pair of rollers, cylinders, or cones, or some such movements, which being twined around by their motion, draws in the new mass of wool or cotton to be spun, in proportion to the velocity given to such rollers, cylinders, or cones; as the prepared mass presses regularly through or betwixt these rollers, cylinders, or cones, others, moving proportionally faster than the first, draw the rope, thread, or sliver to any degree of fineness"—a design that he improved with a new patent in 1758.

The following decade, an even larger step was taken. At almost exactly the same time that Matthew Boulton was creating the Soho Manufactory, Henry Cort was building his puddling furnace, Joseph Black was investigating the properties of latent heat, and James Watt was repairing Glasgow University's broken model of a Newcomen engine, a Lancashire weaver had his own Gestalt moment.*

* Or possibly not. See Thomas Highs's version, below.

The year of James Hargreaves's inspiration is a little vague—his daughter dated it to 1766—but not its character. While visiting a friend, Hargreaves observed a spinning wheel that had been knocked down; with the wheel and spindle in a vertical position, rather than their then-traditional horizontal one, they continued to revolve. In a flash, Hargreaves imagined a line of spindles, upright and side by side, spinning multiple threads simultaneously.

Nearly fifty years later, the first description of the spinning jenny ("jenny" is a dialect term for "engine" in Lancashire) appeared in the September 1807 issue of *The Athenaeum,* in which readers learned that the first one was made "almost wholly with a pocket knife. It contained eight spindles, and the clasp by which the thread was drawn out was the stalk of a briar split in two." The result is not just a romantic tale; the jenny immediately delivered an eightfold increase of the amount of yarn that a single spinner could produce. Just as immediately, the machine had customers—and enemies. Hargreaves's daughter recalled that fearful hand spinners "came to our house and burnt the frame work of twenty new machines."

In June 1770, Hargreaves drafted an application for a patent (number 962) which read, "much application and many trials [produced] a method of making a wheel or engine of an entire new construction that will spin, draw, and twist sixteen or more threads at one time by a turn or motion of the hand and the draw of the other." Those "many trials," and, more significantly, the delay between the first (and very public) sales of the jenny and receipt of legal protection for its design, made for some serious problems. Lancashire's cotton manufacturers* had been using the jenny for at least two years before Hargreaves got around to patenting it, and, under threat of losing their new best friend, they offered Hargreaves £3,000 for a license, which he evidently refused, seriously overplaying his hand. The fact that he had sold the jenny before patenting it severely limited his patent rights, and he died eight years later, comfortable, but not rich.

* One of them was the future prime minister Robert Peel, whose father had been a partner of Hargreaves.

Hargreaves's experience was telling. Both ends of the cloth-making process—spinning and weaving—were dominated by artisans who defined their own interests in terms rather more complicated than a simple desire for wealth. Prior to the introduction of the jenny, Britain's spinning was performed largely by what we would call independent contractors: the original cottage industrialists, taking raw materials from manufacturers who "put out" for contract the production of finished fabric.

This was efficient—no huge capital expenses for the manufacturer, for example—but it contained within its organization what one might call a moral hazard. Since independent spinners worked for more than one manufacturer, they frequently juggled their contracts, delaying manufacturer number one in order to meet an order for number two. At its worst, this meant taking one manufacturer's raw material and using it to produce goods for another, hiding the choice by making a flimsier yarn for both.

The other half of the textile "industry" was no different. Like the lilies of the field, Britain's weavers did not spin, nor did they toil, at least not more than necessary. They were proud artisans who not only wanted to control their work, but also were famously unwilling to work too hard at it. In the words of historian David Landes, "Weavers typically rested and played long, well into the week, then worked hard toward the end in order to make delivery and collect pay on Saturday. . . . Saturday night was for drinking, Sunday brought more beer and ale." And it didn't end there; one of their more rambunctious traditions was the custom known as "Saint" Monday, an ode to which appeared sometime in the 1780s:

When in due course, SAINT MONDAY wakes the day,
Off to a Purl-house straight they haste away;
Or, at a Gin-shop, ruin's beaten road,
Offer libations to a tippling God. . . .

The attitude of the day suited spinners and weavers a lot better than it did the manufacturers who employed them. The obvious solution to their problems was to observe the artisans as they worked, which meant getting them out of their cottages and into factories. One of the more obdurate rules of economics, however, is that, given their capital demands, factories are preferable to more flexibly "outsourced" labor only if they are more productive.

The place where they became so was the same one where the Lombes built England's first textile factory: on the banks of the River Derwent.

A TOUR GROUP DECIDING to engage a boat for a day trip along the Derwent might be forgiven for thinking that they had seen England change from a feudal to an industrial economy between morning and evening. Chatsworth Park, on the banks of the Derwent in Derbyshire's Peak District, is one of the greatest estates in Britain; from the sixteenth century on, Chatsworth House was only one of the many homes of the Dukes of Devonshire, built at the center of the sine qua non of medieval wealth: thousands of acres of productive agricultural land cultivated by tenant farmers. Downstream from Chatsworth Park's legendary gardens, the Derwent flows under a bridge even older than the dukedom, past the village of Darley Dale whose place in history marks another sort of wealth: Darley Dale was the home of Henry Maudslay's assistant, Joseph Whitworth, who established the standard for English screw threads. From there, the river takes a series of U-shaped oxbow turns through a deep set of limestone cliffs and picks up some tributary streams, emerging into a valley with heavy vegetation on both banks, broken, about five miles south of Darley Dale, by a onetime factory now housing a museum and a shopping mall, a sprawling red brick behemoth known as Masson Mills.

Masson Mills, and the Derwent Valley in which it was built, are, like the Iron Gorge foundries, recognized by the United Nations as an official World Heritage site. The reasons given by UNESCO read as follows: "The Derwent Valley saw the birth of the factory system, when new types of building were erected to house the new technology for spinning cotton. . . . In the Derwent Valley for the first time there was large-scale industrial production in a hitherto rural landscape. The need to provide housing and other facilities for workers and managers resulted in the creation of the first modern industrial towns." The system, the buildings, the transformation of the landscape, and even the towns arose on the banks of the Derwent almost entirely because of the efforts of one man: the brilliant, and exceedingly controversial, Richard Arkwright.

ARKWRIGHT WAS BORN IN Lancashire in 1732, to a family at the less prosperous end of the artisan scale—saddlers, tailors, shoemakers, and the like—who apprenticed him at age twelve to a slightly atypical trade. For six years, Arkwright studied the craft of barbering, though it didn't take him long to realize that he could make considerably more money making hair than removing it. For nearly fifteen years, he was a maker of wigs, recalling, "I was a barber, but I have left it off, and I and another are going up and down the country buying hair and can make more of it."

No doubt he found the trade in hair pleasant enough, at least until 1767, when he met, in a pub, an itinerant clockmaker with the confusing (to historians, anyway) name of John Kay. This John Kay had nothing to do with flying shuttles, but he did have an interest in the other side of clothmaking, and he boasted to his new drinking companion, just as he collapsed over his last drink, that he could build a machine that would spin cotton on rollers. As both men later recalled, Kay woke up to find the onetime wig maker looming over him demanding a small model as proof of the clockmaker's boasts.

As would be subsequently revealed, Kay had invented the new spinning machine in much the same way that John Lombe had invented the silk mill. Given the rather fluid attitudes of the day concerning intellectual property, it's probably too much to say that he stole the design, but he certainly borrowed it, from a Lancashire reed maker and weaver named Thomas Highs, who may even have a claim on the invention of the spinning jenny (Highs's

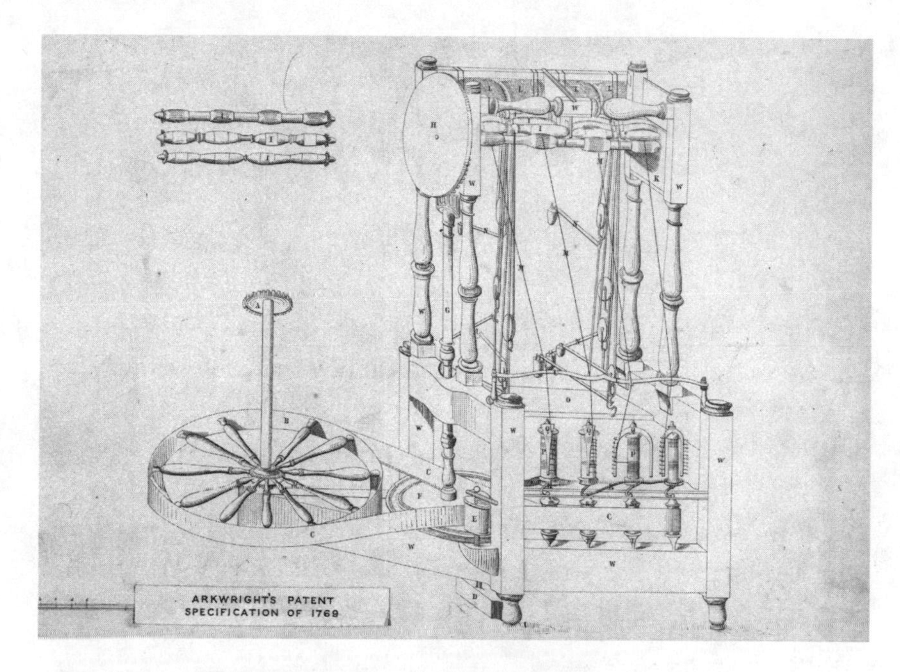

Fig. 6: This is the diagram that accompanied Arkwright's patent application, which became the 931st patent awarded by Britain, in July 1769. The rollers, on the right, teased out the fibers at different speeds, and were then given a helical twist by the wheel, on the left. The original power source was a horse mill driving a vertical shaft attached to a pulley; a belt, in turn, transferred the motion from the pulley to the spindles. *Science Museum / Science & Society Picture Library*

daughter, Jane, always maintained that it was named for her). Whatever his contribution to the jenny, he was clearly responsible for the design of the machine that Kay reproduced—from memory—for Arkwright, since two years before, Highs had hired the clockmaker to turn his wooden model into a working machine made of iron.

The Highs design did have one unique and important feature. While Paul's spinning machine had only one set of rollers, Highs's had two, with the second rotating five times as fast as the first. The second rollers therefore stretched the original thread fivefold before the jennylike bobbin-and-flyer gave it the needed twist, producing thread that was both longer and stronger than could be produced by either hand or jenny.

This was huge. English cotton was finally strong enough and

long enough to be used not merely for weft but as warp thread, replacing both the more expensive linen and the Indian cotton that had been banned by the Calico Acts. Highs had invented the machine that would, more than any other, create Britain's cotton industry.

He did not, however, patent his creation, perhaps because he did not yet see the profit to be made from it. Richard Arkwright, however, did. In 1768, after keeping Kay sequestered in Nottingham for a full year, he applied for a patent, which was awarded in 1769, the same year that James Watt patented his separate condenser. Two years later, Arkwright and his two partners, Samuel Need and Jedediah Strutt, started building, at Cromford in the Derwent Valley, the first factory to use the new spinning machine, which he named the water frame—"water" because Arkwright's cotton was to be spun by harnessing the current of the River Derwent.

It is scarcely surprising that Arkwright's cotton mill was dependent on waterpower. Though Newcomen engines were by then familiar sights at mine shafts all over England, Britain's waterwheels produced at least ten times as much power as steam did, and the source of rotary motion for the rollers in Arkwright's water frames was a wheel set below a millrace off the Derwent River, harnessing not only water flow but gravity to deliver a nominal ten horsepower. It opened in 1771 and a year later was already a huge success, based on the enormous labor savings of the mill (as Arkwright himself wrote, "wee [*sic*] shall not want ⅕ of the Hands I First Expect.d"); the high quality of the cotton, which approached silk in its smoothness; and, as was predictably the case, Jedediah Strutt's successful lobbying of Parliament to reduce the taxes on British-made cotton.

Success can cement the relations between partners; such was the story of Matthew Boulton and James Watt. It can also corrode them. In 1775, the partnership of Arkwright, Strutt, and Need was showing some signs of rot. That was the year that Arkwright applied for, and received, an entirely new series of patents for machines that could, sequentially, card and comb raw cotton, draw it into thread, and twist it into yarn, intended to centralize every aspect of the manufacturing process. Significantly, the 1775 patents were in Arkwright's name only, unlike the original 1769 water frame patent, which included the two partners (though not Kay, who had been formally apprenticed to Arkwright and subsequently left after much conflict with his master). The partnership did not survive, though both Strutt and Need were well compensated for their few years' investment when Arkwright bought out their shares in the Cromford mill.

In 1774, another Lancashire inventor, Samuel Crompton, had produced the first machine to integrate a spindle carriage with a weaving frame, and could thus take raw material in one end and produce cloth at the other. But Crompton, either out of perversity or lack of ambition, never patented his invention, apparently being more interested in the writings of the Swedish scientist, inventor, and mystic Emanuel Swedenborg. As a result, though Richard Arkwright paid a small fee to see the machine demonstrated in 1780, he had no obligation to pay subsequent royalties to its inventor, and almost immediately he incorporated Crompton's invention—the first "mule," so called because of its mixed parentage—into his factories, of which the most spectacular was Masson Mills.

If the Derwent River mills weren't already, Crompton's mule made them the most productive, if not the most enlightened, in the world, running two twelve-hour shifts daily. The machines still required adult strength and skill to operate, but numerous tasks, including gleaning the unused cotton, gave employment (if that is the right word) to hundreds of children as young as six. When the brilliant engineer and indefatigable improver John Smeaton demonstrated that breastshot waterwheels, which caught the flow of water as it fell from a millrace, were more efficient than the undershot wheels that were turned only by a river's current, Arkwright changed his power source within months.

He was, partly because of his success with waterpower, suspicious of steam, about which he displayed an atypical indecisiveness, inquiring about a Boulton & Watt engine as early as 1777 but waiting until 1790 to order one. While he preferred waterpower, he was scarcely dogmatic about it; his Nottingham factory continued to use horse-powered wheels long after he shifted to water, partly as an experiment in calculating his power needs; he could scarcely add to or subtract from a river's flow in the same way that he could add or subtract horses on their wheel. Horse power, in addition (and again unlike waterpower), was scalable: As a factory grew, it was easier to augment horse power than waterpower.

By the 1780s, however, Arkwright finally caught an enthusiasm for returning engines, the kind that use steam pumps to transport water to a higher elevation, which allowed gravity flow to operate a waterwheel. In 1781, Smeaton evaluated the potential of mills worked by steam directly versus those mediated by water and wrote that "no motion can ever act perfectly steady and equal in producing a circular motion, like the regular efflux of water turning in a waterwheel." Arkwright took him at his word. He built the Shudehill mill, in Manchester, which used two Newcomen-type steam engines, consuming five tons of coal daily, to pump water to a reservoir from which it could drive a waterwheel thirty feet in diameter and eight feet wide, recycling the water all day long; the "earliest steam-powered cotton spinning mill was driven by the earliest successful type of steam engine."

By the time he built Shudehill, Richard Arkwright employed at least five thousand people and his estimated net worth was somewhere north of £200,000. That was also the year in which he decided to take his winning streak to court, filing suit to protect his rights in the underlying 1769 patent, which was due to expire in 1783, along with his royalties on the fundamental machine used in cotton spinning. Arkwright had consistently set those royalties very high, partly to protect his own manufacturing businesses, and as a result had no shortage of infringers; in 1781, Arkwright sued nine of them, and the court found for him eight times. Unsatisfied, he kept up the pressure for another four years, even after the expiration of his first and most important patent.

In February 1785, Arkwright filed suit against his Derbyshire neighbor "Mad Peter" Nightingale* to finally recover the carding

* This nickname for Nightingale, great-uncle of Florence, was apparently earned by his daredevil horseback riding.

portion of his 1775 patent, and although he secured a finding of infringement, Arkwright had finally overreached. In May 1785, the Crown, under pressure from Arkwright's competitors, filed a writ of *scire facias,* using an archaic legal doctrine that required a sheriff to notify a party that his right was questioned and had to be defended. By placing the burden of proof on Arkwright rather than his accusers, his competitors in the cotton industry, who had invested hundreds of thousands of pounds in machinery that they understandably wanted to be able to use without permission from Arkwright, had stumbled on a powerful weapon.

The trial of *Rex v. Arkwright,* which was heard at Westminster Hall in June 1785, was the result. The original dubiousness of Arkwright's "invention" now came back to haunt him, as Highs and Kay, and even James Hargreaves's widow, all appeared as witnesses against him, with Kay going on record as saying "he never would have had the rollers but through me." Arkwright, during his own testimony, said, "if any man has found out a thing, and begun a thing, and does not go forwards . . . another man has the right to take it up, and get a patent for it."

The final ruling, by Chief Justice Buller, found against Arkwright on three separate grounds: that the 1775 patent was not novel (that it essentially restated the 1769 patent, in an attempt to extend it); that it included elements not invented by Arkwright; and that it was insufficiently specific. On November 14, 1785, the Court of King's Bench vacated four of Arkwright's patents.

He was enraged, but hardly impoverished, by the ruling. The real impact, however, was the unique public forum it offered Britain on the subject of patent, invention, and the new world that they had created. Though the public attacks on Arkwright's behavior were vicious (as were the courtroom tactics: King's Counsel Edward Bearcroft pointed to Arkwright and declared, "There sits the thief!"), the actual decision against him was based on the technical grounds that his original specification was too vague. Though the last piece of the decision was the least newsworthy—the broadsheet distributed immediately after the trial, which crowed that "the old Fox is at last caught by his own beard in his own trap," made no mention of it—it was by far the most significant.

By failing to describe the invention adequately, Arkwright's patent application had, in essence, broken the bargain that granted patents to inventors in return for their making public the useful knowledge inherent in them. This part of the ruling would draw the attention of a number of other inventors, including James Watt.

Watt had been drawn into the Arkwright litigation in January 1785, at the behest of Boulton, who received a letter from Erasmus Darwin that said in part, "If yourself or Mr. Watt think as I do on this affair, & that your own interest, pray give me a line that I may advise Mr. Arkwright to apply to you." Watt replied, "Though I do not love Arkwright, I don't like the precedent of setting aside patents through default of specification. I fear for our own. The specification is not perfect according to the rules lately laid down by the judges. Nevertheless, it cannot be said that we [Boulton & Watt] have hid our candle under a bushel. We have taught all men to erect our engines, and are likely to suffer for our pains. . . . I begin to have little faith in patents; for according to the enterprising genius of the present age, no man can have a profitable patent but it will be pecked at. . . ."

Watt did more than simply offer testimony in the Nightingale trial. After the final decision in November 1785, Josiah Wedgwood, like Watt and Boulton a member of the Lunar Society of Birmingham and himself a successful industrialist, wrote to Watt, "I have visited Mr. Arkwright several times and find him much more conversible than I expected. . . . I told him you were considering the subject of patents, and you two geniuses may probably strike out some new lights together which neither of you might think of separately." Wedgwood proved prescient; together, Watt and Arkwright wrote a manuscript entitled "Heads of a Bill to explain and amend the laws relative to Letters Patent and grants of privileges for new Inventions," essentially a reworking of Coke's Statute of 1623 that had created England's first patent law. In addition to its policy prescriptions, which were largely an unsuccessful argument against the requirement that patent applications be as specific as possible, the manuscript offered a remarkable insight into Watt's perspective on the life of the inventor, who should, in Watt's own (perhaps inadvertently revealing) words, "be considered an Infant, who cannot guard his own Rights":

> An engineer's life without patent is not worthwhile . . . few men of ingenuity make fortunes . . . without suffering to think seriously whether the article he manufactures might, or might not, be Improved. The man of ingenuity in order to succeed . . . must seclude himself from Society, he must devote the whole powers of his mind to that one object, he must persevere in spite of the many fruitless experiments he makes, and he must apply money to the expenses of these experiments, which strict Prudence would dedicate to other purposes. By seclusion from the world he becomes ignorant of its manners, and unable to grapple with the more artful tradesman, who has applied the powers of *his* mind, not to the improvement of the commodity he deals in, but to the means of buying cheap and selling dear, or to the still less laudable purpose of oppressing such ingenious work men as their ill fate may have thrown into his power.

At no earlier time or other place in human history could Watt's argument—"patents create a great and profitable trade . . . to the immense emolument of the state," which should therefore grant patents "not as the price of a secret, but as rewards to men of merit for their ingenuity"—have even been comprehensible.* Combin-

* It's not fully comprehensible even today, since it depends on a principle that remains problematic: the belief that a direct line can be drawn from a single invention to a single inventor. In the 1920s, the historians William Ogburn and Dorothy Thomas first documented the notion of "multiples"—simultaneous discovery by different people, which occurred with the telephone (Bell and Elisha Gray), thermometer (six different inventors), steamboat (Fulton, Jouffroy, Stevens, etc.), and calculus (Newton and Leibniz). Robert K. Merton wrote an essay on scientific discovery in the 1960s suggesting that the more gifted the scientist, the more likely that his discoveries will be multiply discovered, thus inspiring the statistician Stephen Stigler to formulate Stigler's Law: No scientific discovery is ever named after its original discoverer.

ing Locke's seventeenth-century doctrine of natural rights in one's intellectual labor with eighteenth-century utilitarianism, it was, literally, revolutionary.

Arkwright died a wealthy man in 1792; within a few decades, biographers would describe him as a fraud who had, in one biographer's words, "possessed unwearied zeal and patience in obtaining the discoveries of others." A few years later, others would defend him; one described him as "a man of Napoleonic nerve and ambition." By the 1840s, the retrospectively apparent fact that Arkwright, and men like him, had made Britain a world-straddling power pretty much guaranteed a measure of florid, if backhanded, hero worship. The greatest hero-worshipper of them all, Thomas Carlyle, described Arkwright as

> A plain, almost gross, bag-cheeked, potbellied, much enduring, much inventing man and barber. . . . French Revolutions were a-brewing: to resist the same in any measure, imperial Kaisers were impotent without the cotton and cloth of England, and it was this man that had to give England the power of cotton. . . . It is said ideas produce revolutions, and truly they do; not spiritual ideas only, but even mechanical. In this clanging clashing universal Sword-dance which the European world now dances for the last half-century, Voltaire is but one choragus [leader of a movement, from the old Greek word for the sponsor of a chorus] where Richard Arkwright is another.

This touches on, but misses, the importance of the part played by Arkwright in the birth of self-sustaining industrialization. Sooner or later, the cycle of innovation needed to provide goods not merely for Britain's manufacturers and traders—what a modern analysis would call business-to-business commerce—but the nation's consumers. The typical eighteenth-century British household could scarcely buy cannon, or wooden pulley blocks for sailing ships, or—except for home heating—even much coal. But they could, and did, buy clothes. Arkwright was not a great inventor, but he was a visionary, who saw, better than any man alive, how to convert useful knowledge into cotton apparel and ultimately into wealth: for himself, and for Britain.

IN ADDITION TO MAKING cloth and inspiring innovation, textile manufacturing produced conflict. More than the mining of coal, or the making of iron, or even the grinding of grain, it exposed the great social clash of the day: on the one hand, the power of sustained innovation, fueled by ever-increasing wealth; on the other, five centuries of traditional expertise controlled by militant and well-organized artisans. None of them were more militant, or better organized, than Britain's spinners and weavers.

Spinning first. The art of spinning is largely a matter of coordinating several processes simultaneously so that the fiber is under constant tension. Since it is elastic, the amount of tension applied to it while it is wound can result in yarn that is inconsistent in quality from one end to the other. In the first spinning machines, the operator had to simultaneously shape the winding and turn the spindles at precisely the same rate, so as to wind up the yarn—the term of art is "winding the cop"—without either stretching the yarn or allowing it to go slack. The craft was difficult enough that spinners became not only indispensable to the process, but highly protective of their place in it, exhibiting all the rent-seeking mania of a medieval guild. Along the way they transformed themselves from independent contractors into the nation's most powerful and highly organized craft union. At one union meeting, a spinner argued violently against allowing "piecers" (the subordinates on the spinning line, who tie together threads when they break) to actually put up a cop of cotton yarn unless he was "a son, brother, or orphan nephew." In the industry's Lancashire heartland, mule spinners developed work rules in 1780 that remained in force until the 1960s, and partly in consequence, the new and improved ring-spinning machines, invented by the American John Thorp in 1828, which operated continuously and twisted fibers into yarn by attaching them to a rotating ring, didn't catch on in Britain until the end of the nineteenth century.

As with spinning, so with weaving. Edmund Cartwright, a onetime Church of England minister and "the last of the great inventors who belong to the craft period," built the first power loom in 1785, inspired by the need to keep up with the great surpluses of yarn being produced by Arkwright's factories.* As Cartwright later recalled,

> as soon as Arkwright's patent expired, so many mills would be erected and so much cotton spun that hands would never be found to weave it. . . . It struck me that as plain weaving can only be three movements which were to follow each other in succession, there would be little difficulty in producing them and repeating them. Full of these ideas I immediately employed a carpenter and smith to carry them into effect. As soon as the machine was finished, I got a weaver to put in a warp which was of such material as sail cloths are usually made of. To my great delight, a piece of cloth, such as it was, was the product.

Cartwright's initial design, for which he received a patent in 1785, was ingenious, but not yet practical, because it failed to solve the feedback problem inherent in the nature of mechanizing the shuttle, which was, in the language of engineering, "negatively driven." That is, it was driven first one way, then the other, which meant that it could not be allowed to rebound and so slacken the weave. The task of maintaining the constant tension needed to keep each thread of the warp the same length was hugely difficult to mechanize, which was why weaving remained a handcraft longer than any other step in textile manufacturing. Among other

* He was also, apparently, convinced of the practicality of such a machine by the success of the "Mechanical Turk," a supposed chess-playing robot that had mystified all of Europe and which had not yet been revealed as one of the era's great hoaxes: a hollow figurine concealing a human operator. Inventors are sometimes beneficiaries of their own ignorance.

things, Cartwright needed some way to control variations in speed, since a too-slow shuttle wouldn't travel the entire width of the loom, and a too-fast one would bounce back, with disastrous results.

It took two years, and three more patents, before the Reverend Cartwright's loom was ready for commercial application, but in April 1787, the new and improved version, with its frame now horizontal rather than vertical, and with each warp thread attached to a separate bobbin, was complete.

Cartwright constructed twenty looms using his design and put them to work in a weaving "shed" in Doncaster. He further agreed to license the design to a cotton manufacturer named Robert Grimshaw, who started building five hundred Cartwright looms at a new mill in Manchester in the spring of 1792. By summertime, only a few dozen had been built and installed, but that was enough to provoke Manchester's weavers, who accurately saw the threat they represented. Whether their anger flamed hot enough to burn down Grimshaw's mill remains unknown, but something certainly did: In March 1792, after a series of anonymous threats, the mill was destroyed.

Cartwright's power looms were not the first textile machines to be attacked, and they would not be the last.

SIR ISAAC NEWTON'S THIRD law of motion states that every action is paired with an equal and opposite reaction. The "equal and opposite" reaction to the industrialization of the textile industry—and, by extension, all industrialization—is widely, though vaguely, known as Luddism.

Resistance to the mechanization of the traditional crafts of spinning and weaving had been around for two centuries before anyone heard of Luddism, or Luddites. In 1551 Parliament passed legislation prohibiting mechanical gig mills, used to raise the nap on wool, and William Lee, inventor of an early knitting frame, was forced by the hosier's guild to leave England in 1589. More often, hostility to machinery made itself known not in the form of writs and laws but crowbars and clubs. In 1675, weavers in Spitalfields attacked engines (not, of course, steam-powered) able to multiply the efforts of a single worker. Not only was Richard Hargreaves's original spinning jenny destroyed in 1767, but so also was his new and improved version in 1769.

Nor was the phenomenon exclusively British. Machine breaking in France was at least as frequent, and probably even more consequential, though it can be hard to tease out whether the phenomenon contributed to, or was a symptom of, some of the uglier aspects of the French Revolution. Normandy in particular, which was not only close to England but the most "English" region of France, was the site of dozens of incidents in 1789 alone. In July, hundreds of spinning jennys were destroyed, along with a French version of Arkwright's water frame. In October, an attorney in Rouen applauded the destruction of "the machines used in cotton-spinning that have deprived many workers of their jobs." In Troyes, spinners rioted, killing the mayor and mutilating his body because "he had favored machines." The carders of Lille destroyed machines in 1790; in 1791, the spinning jennies of Roanne were hacked up and burned. By 1796, administrators in the Department of the Somme were complaining, it turns out presciently, that the "prejudice against machinery has led the commercial classes . . . to abandon their interest in the cotton industry."

The Luddite version of machine breaking—what the historian E. J. Hobsbawm called "collective bargaining by riot"—was the product of half a dozen different but related historical threads. One was surely the Napoleonic Wars, which had been under way more or less continuously for more than fifteen years by the time of the first Luddite activity. The war economy had affected the textile industry of the Midlands no less than the shipbuilders of Portsmouth, first with dramatic increases in demand for sailcloth and uniforms, and then—as Napoleon's so-called "Continental System" restricted British trade with the Americas and Europe—with equally dramatic decreases in exports, which fell by nearly a third from 1810 to 1811.

Bad enough to be a manufacturer in such times; far worse to be a laborer. Handloom weavers had been earning nearly twenty shillings a week in the 1790s; twenty years later, mostly because of the large number of new entrants to the industry, they were now earning less than ten. Factory workers were paid better, but the conditions in which they worked could be much worse: lung-destroying cotton dust everywhere, and noise so loud that workers were obliged to invent a method of lip-reading (known in Lancashire as "mee-mawing").

Many were nonetheless driven to factory work by the dramatic increase in the amount of rural land removed from the commons—the so-called "enclosure" movement by which more than six million acres of fields, meadows, and forests representing more than half of all the land then in cultivation in England were hedged, fenced, and turned into private property between 1770 and 1830. Enclosure was bad enough; in combination with war-fueled inflation, it doubled the price of food: a loaf of bread that had cost ninepence in 1800, by 1810 carried a price of a shilling and fivepence. The effective increase, to a handloom weaver who had seen his income halved in the same decade, was even more onerous, from 4 percent of his weekly wage to more than 14 percent.

Resentment and hunger among Britain's weavers made for an explosive mixture. The first shots of the "rebellion," however, were fired, not by weavers of broadloom but by an even more militant subset of textile artisans: Nottingham's stocking knitters.

The technique of knotting and looping a single length of yarn into a continuous fabric is a fairly new technique, at least as compared to weaving, but versions still date back several thousand years. Evidence of the earliest, the single-needle craft known retrospectively by the Norse term *nalbinding,* has been found in the third-century CE Syrian city of Dura-Europos, but the use of double needles to pull one knot through another didn't replace the far more difficult technique until the early Middle Ages. Double-needle knitting is not only faster, but it can, by selective choice of dropped stitches, curve in three dimensions, and it is therefore a highly attractive method for producing garments that need to be form-fitting. For most of human history, this meant extremities: gloves for hands, and (especially) stockings for feet. The skill required to produce a knitted sock by hand was great enough, and the investment in training so expensive, that stockingers were even more opposed to mechanization than either spinners or weavers.

This did not, of course, eliminate the urge to invent. The stocking frame, the world's first knitting machine, was designed and built in 1589 by William Lee of Nottingham, a onetime curate who twice attempted to secure patent protection, failing both times. The lack of a patent took royalties out of Lee's pocket but did nothing to stall the widespread adoption of the machine, which increased speed from a hundred stitches per minute to a thousand. This demanded a response from the stocking makers; in 1657, during the Protectorate, the London Company of Framework Knitters persuaded Oliver Cromwell to grant them a charter, and thus effective control over the production of knitted fabric throughout England. Sixty years later, disputes between the guilds of London and Nottingham ended with the latter independent of its parent guild and home to lots of new stockingers.* In the late 1770s, they petitioned Parliament to formalize their exclusive ownership of their craft with a law entitled "the Art and Mystery of Framework Knitting," and when it failed to pass, they rioted in Nottingham.

The stockingers' fierce defense of their prerogatives thus had a long list of precedents on March 11, 1811, when the first shots of the Luddite "rebellion" were fired. By then, Nottingham alone probably had nine thousand stocking frames, Leicestershire and Derbyshire another eleven thousand—and fewer than half were in use. Their owner-operators had become victims of both competition from factory "cut-ups" (stockings sewn together from two or more knit pieces, which had the benefit of being easier and cheaper to make, though far less sturdy) and the vagaries of Regency fashion;

* And lots of new stockings. Knitting anything but wool or silk was a pretty daunting task until 1758, when Jedediah Strutt—Richard Arkwright's partner—patented the Derby Rib, which alternated two stitches, one the reverse of the other, and so made the production of ribbed cotton stockings as practical as that of silk.

the legendary dandy Beau Brummel, who famously claimed that a frugal man could, with discipline, dress himself for no more than £1,000 a year, preferred trousers to knee socks.

The stockingers began in the town of Arnold, where weaving frames were being used to make cut-ups and, even worse, were being operated by weavers who had not yet completed the seven-year apprenticeship that the law required. They moved next to Nottingham and the weaver-heavy villages surrounding it, attacking virtually every night for weeks, a few dozen men carrying torches and using prybars and hammers to turn wooden frames—and any doors, walls, or windows that surrounded them—into kindling. None of the perpetrators were arrested, much less convicted and punished.

The attacks continued throughout the spring of 1811, and after a brief summertime lull started up again in the fall, by which time nearly one thousand weaving frames had been destroyed (out of the 25,000 to 29,000 then in Nottingham, Leicestershire, and Derbyshire), resulting in damages of between £6,000 and £10,000. That November, a commander using the *nom de sabotage* of Ned Ludd (sometimes Lud)—the name was supposedly derived from an apprentice to a Leicester stockinger named Ned Ludham whose reaction to a reprimand was to hammer the nearest stocking frame to splinters—led a series of increasingly daring attacks throughout the Midlands. On November 13, a letter to the Home Office demanded action against the "2000 men, many of them armed, [who] were riotously traversing the County of Nottingham."

By December 1811, rioters appeared in the cotton manufacturing capital of Manchester, where Luddites smashed both weaving and spinning machinery. Because Manchester was further down the path to industrialization, and therefore housed such machines in large factories as opposed to small shops, the destruction demanded larger, and better organized, mobs. Because the communities they targeted were likewise better protected—Manchester alone had more than three thousand men serving as constables or members of the city's night watch—it also put more of them at risk of capture, and by 1812, dozens of Luddites were on trial. Most were acquitted, but all were required to take loyalty oaths, and those at risk of punishment were granted royal pardons, though only under condition that they renounce Luddism and reaffirm their loyalty to the Crown on pain of death.

By 1812, however, the riots had started to inspire other disaffected laborers. In January, the West Riding of Yorkshire was subject to regular attacks by groups of "croppers" (men who used fifty-pound hand shears to cut the nap from woolen cloth, thus making it smooth) in fear for their jobs by the introduction of yet other new inventions: the once-banned gig mill, which raised the nap of the wool so that it could be sheared; and the complete shearing frame, which made a slightly inferior article, but could be operated by relatively unskilled workers.

On January 1, the Framework Knitters issued a document that declared, among other things,

> Whereas by the charter granted by our late sovereign Lord Charles II by the Grace of God King of Great Britain France and Ireland, the framework knitters are empowered to break and destroy all frames and engines that fabricate articles in a fraudulent and deceitful manner and to destroy all framework knitters' goods whatsoever that are so made and whereas a number of deceitful unprincipled and intriguing persons did attain an Act to be passed in the 28th year of our present sovereign Lord George III whereby it was enacted that persons entering by force into any house shop or place to break or destroy frames should be adjudged guilty of felony and as we are fully convinced that such Act was obtained in the most fraudulent interested and electioneering manner and that the honourable the Parliament of Great Britain was deceived as to the motives and intentions of the persons who obtained such Act we therefore the framework knitters do hereby declare the aforesaid Act to be null and void. . . .
>
> Given under my hand this first day of January 1812.
>
> God protect the Trade.
>
> *Ned Lud's Office—Sherwood Forest*

The choice of words is revealing. The knitters believed themselves to be not merely injured economically, but victims of fraud and deceit. This made them not only self-interested but self-righteous, and through the spring of 1812, attacks grew more and more violent, with total damage estimated at £100,000 and at least a dozen deaths, almost all of them Luddites shot by horsemen from the Scots Greys, an army troop quartered nearby. In self-defense, the Luddites began targeting armories in order to equip themselves with firearms and ammunition; more alarming to the national government, the mobs had adopted Jacobin vernacular and costume, including the red flag, the *drapeau rouge,* of the Revolution. The national government was ready to react, or, more precisely, overreact. In February 1812, frame breaking was made a capital offense, and twelve thousand soldiers—roughly the number of British troops the future Duke of Wellington had led into battle against the French in Portugal four years earlier—deployed to enforce it. During a single Luddite attack on a Lancashire steam loom on April 18, five were killed and eighteen wounded. Hundreds were transported to Australia, and even more imprisoned. On April 28, a group of Luddites led by the onetime cropper George Mellor attacked the Rawfolds Mill and killed William Horsfall, its owner. Newspapers were reporting not just a local insurrection but a national rebellion.

Much of it was exaggeration. In July 1812, another letter to the Home Office, this from Earl Fitzwilliam, Lord Lieutenant of the West Riding, described a somewhat less frantic scene:

> I do not mean to say, that parties of Luddites have not been met travelling from place to place, and perhaps marshalled in some degree of order, but that there is no evidence whatever, that any one person has yet established the fact of their having been assembled and drilling in a military way—as far as negative evidence can go, I think, the contrary seems established.

The Luddite legend has survived for centuries in part because of the appeal of a romantic brotherhood, a secret society complete with blood oaths: "I, ___, of my own free will and accord do hereby promise, and swear that I will never reveal any of the names of any one of this secret committee, under the penalty of being sent out of this world by the first brother that may meet me, I furthermore do swear, that I will pursue with unceasing vengeance any traitor or traitors . . ." and even secret signals and passwords:

> You must raise your right hand over your right eye if there be another Luddite in company he will raise his left hand over his left eye—then you must raise the forefinger of your right hand to the right side of your mouth—the other will raise the little finger of his left hand to the left side of his mouth and will say What are you? The answer, Determined—he will say, What for? Your answer, Free Liberty—then he will converse with you and tell you anything he knows. . . .

Though they would not have used the terms, the Luddites were on one side of a newly violent debate about the relationship between labor and property. Opposing them was the newfangled notion that *ideas* were property; the Luddites argued (with crowbars and torches) that their *skills* were property. The right of men to enjoy the fruits of their labor gave them license to defend the free exercise of those skills in exactly the same way that they might defend their houses.

The Luddite rebellion failed for the most obvious reason: an enormous disparity in military power, power that the national government was, eventually, willing to bring to bear. The Luddite idea lost the historical battle—"Luddite" is not, in most of the contemporary world, used as anything but an insult—because its thesis, which might be abbreviated as "property equals labor plus skill," was less attractive than the idea that property equals labor plus ideas. The victory of the latter was decided not by argument, but economics: it produced more wealth, not just for individuals, but for an entire nation. Over time, the patents of Lombe, Kay,

Hargreaves, and Arkwright not only became public property but attracted competing and superior inventions. In 1813, there were 2,400 power looms in England; in 1820 there were 12,150, and by 1833 more than 85,000. With the introduction of the iron power loom by Henry Maudslay's onetime assistant Richard Roberts in 1822, a weaver could produce seven pieces of cotton shirting in a week, each twenty-four yards long, while a hand weaver could make only two. Three years later, the same weaver would average twelve weekly, and six years after that, "a steam-loom weaver, from 15 to 20 years of age, assisted by a girl about 12 years of age, attending to four looms, [could] weave eighteen similar pieces in a week; some can weave twenty." During the century and a half that followed the Calico Acts, the productivity of the cotton industry increased fourteenfold.

We feel real poignancy when we recall the bucolic life (even if we do so through the soft focus of nostalgia) of a country weaver happy in his work skills and content with his life. But those skills, like those of a medieval goldsmith or an ancient carpenter, could not, by their very nature, reproduce themselves outside the closed community of the initiates. One lesson of the Luddite rebellion specifically, and the Industrial Revolution generally, is that maintaining the prosperity of those closed communities—their pride in workmanship as well as their economic well-being—can only be paid for by those outside the communities: by society at large. A great artisan can make a family prosperous; a great inventor can enrich an entire nation.

CHAPTER ELEVEN

WEALTH OF NATIONS

concerning Malthusian traps and escapes; spillovers and residuals; the uneasy relationship between population growth and innovation; and the limitations of Chinese emperors, Dutch bankers, and French revolutionaries

IT TOOK ABOUT SIX hundred years for the publishing industry to get from Johann Gutenberg to the book you are now reading.* The most remarkable eight-week stretch in all of those six centuries fell between January and March 1776—a year overloaded with significant dates. On January 10, Thomas Paine published the pamphlet *Common Sense* ("Society is produced by our wants, government by our wickedness"). February 17 saw the first volume of Edward Gibbon's *History of the Decline and Fall of the Roman Empire* roll off the presses ("In the second century of the Christian Era, the empire of Rome comprehended the fairest part of the earth . . ."). And on March 9, a former University of Glasgow colleague of Joseph Black published *An Inquiry into the Nature and Causes of the Wealth of Nations.*

Standing in London's Science Museum, in front of *Rocket*

* This is a particularly strong argument against a belief in progress.

and surrounded by models of Thomas Newcomen's beam engine, Joseph Bramah's challenge lock, and James Watt's separate condenser, it takes very little imagination to see connections with iron foundries, coal mines, and even cotton fields. The road back to Adam Smith requires more thought, but is just as important, and as enlightening.

Smith's book, like Darwin's *Origin of Species,* was revolutionary in its impact, immediately and permanently, though both are far more frequently cited than read. In particular, *Wealth of Nations* demonstrates that Britain's eighteenth-century transformation—the schoolboy's "wave of gadgets"—was a revolution not merely in technology but in commerce. The founding text of economic science is staggering in its range, with disquisitions on the origin of money, the nature of commodities, interest rates, profitability, the mechanics of trade, bank interest, taxation, public debt, agriculture, and manufacturing. It devotes thousands of words to histories of Europe's towns from the end of the Roman Empire to the present, and of colonial policy from the time of ancient Greece. It is telling, therefore, that Smith decided to open his magnum opus with a section entirely devoted to the "causes of improvement in the productive powers of labour [and] in the skill, dexterity, and judgment with which labour is applied in any nation." This was a remarkable bit of insight, the application of Locke's labor theory of value to national policy.

Smith argued that two conditions were necessary for labor to produce the maximum amount of wealth: perfect competition among sellers—everyone pursuing his or her selfish interest, the famous "invisible hand"—and the complete freedom of buyers to substitute one commodity for another. Under such ideal circumstances (Smith was not the first economist, but he was probably the first to "assume a can opener," i.e. perfect conditions, in a model), specialization, or division of labor, was inevitable. Ten men could each bake their own bread, weave their own cloth, and build their own houses, but if one became a baker, another a weaver, and a third a builder, the result would be more food, clothing, bricks . . . and trade.

Smith's theorems did a spectacular job of explaining the self-regulating character of a free market, in which prices and profits are forced by competition to the lowest possible level.* They inspired David Ricardo's exposition, in 1817, of the principle of diminishing returns: his argument that the growth of the first decades of industrialization was certain to level off, as each successive improvement produced smaller results. Helped along by the inflation in food prices caused by the Napoleonic Wars, they even set the stage for Thomas Malthus's *Essay on the Principle of Population,* with its famous argument that population always grows geometrically, food production arithmetically.

What they didn't do was explain how wealth, profit, *and* competition can all grow over time. In short, it didn't explain the two centuries of growth that were beginning just as *Wealth of Nations* was being published. It is in no way a criticism of the book to state that it covered everything *except* the reason the author's own nation was about to get wealthier than any other nation in the history of mankind. The failure is pretty much explained by what is *not* in the book. Despite living in the middle of the biggest explosion of inventive activity ever recorded, and even though his illustration of the advantages of specialization was a factory for making pins, Smith's book hardly mentions the role of the new machines then transforming his world. Next to nothing about waterpower, to say nothing of steam; nothing about the forging of iron, and his few paragraphs about the textile revolution are mostly an argument for restricting the export of spinning machines. His pin factory, it turns out, was only a metaphor; he never set foot inside one.

Nor did he show any understanding of Darby's furnace, or Arkwright's water frame, or Watt's double-acting engine—none

* David Warsh points out, in his own *Knowledge and the Wealth of Nations* (which has inspired much of this chapter), that Smith's book was the first of only four dominant textbooks the field has ever known; the *only* one until Ricardo's *Principles of Political Economy and Taxation,* which was followed by Alfred Marshall's *Principles of Economics* and eventually by Paul Samuelson's *Economics.* The pattern of successively shorter titles seems finally to be at an end.

of the escape hatches out of humanity's millennium-long Malthusian trap. The efficiencies of specialization are real, and the self-regulating "invisible hand" powerful, but it was the machines, and nothing else, that allowed Britain, and then the world, to finally produce food (or the wealth with which to buy food) faster than it produced mouths to consume it.

A lot more is known about how population increases than how wealth grows. Indeed, the Industrial Revolution was decades old before anyone realized that wealth was growing at all. The first edition of Malthus's *Essay on Population* was published in 1798 and convinced nearly everyone that the hoofbeats of the horsemen of the Apocalypse could already be heard throughout England. In 1817, the English economist David Ricardo predicted that land rents would increase while wages would approach subsistence level, at precisely the moment when British farmland rents per acre started to plummet and the wages of laborers to explode. Partly this was evidence of the limits of accounting with very little data; Britain's first census, inspired by Malthus, wasn't conducted until 1800. But even more it was the lack of a model that carved up overall growth into its constituent parts.

In 1890, another economist, the mathematically trained Cambridge scholar Alfred Marshall, suggested that the century of growth in both income and wealth that began just as Ricardo predicted its opposite was largely due to ideas whose benefits spilled over into the economy soon after they had enriched their creator. An idea—a separate condenser, for example, or a spinning jenny—might be costly for one inventor to develop, but it wasn't long (not even the fourteen years of a patent) before it became de facto public property, and inspired others to improve upon it.

Marshall's "spillovers" were intriguing but remained anecdotal until the 1950s, when the Nobel Prize–winning economist Robert Solow incorporated something very like it into an equation known as the *fundamental equation of growth.* Working from a contemporary economic model that had shown that capital, labor, and land could be substituted for one another—as one component grew more expensive, producers could substitute one for the other—Solow was able to calculate the rate at which average workers increase their output. He found three components of output increase, each one reflecting the key inputs to the national wealth calculus. The first two, land per worker and capital per worker, are, if not easy, at least possible to measure. Except during times of dramatic depopulation, such as the Black Death of the fourteenth century, or extremely large additions to the stock of arable land, as with Europe's discovery of the New World, growth in land per worker has been negligible for centuries, so small that its effect on growth can be eliminated in the simplest calculations. The second component, growth in capital per worker—that is, all the buildings, machinery, tools, and so on—explains only about 24 percent of total growth. However, since the growth in the amount of land and capital per worker together doesn't equal the overall growth rate, a fudge factor must be used, called the *residual:* what's left over.

This also means that the residual, despite the ass-backward way it is calculated, amounts to at least three-quarters of the total increase in economic growth since 1800. That's a big chunk of activity defined by subtracting everything else, a little like a ten-drawer file cabinet with seven drawers marked "Miscellaneous." Solow first assumed that the residual represented increasing efficiency over time, and he incorporated an arbitrary constant to represent the rate of the growth in useful knowledge.

Useful knowledge, in this formulation, is not *all* knowledge. The growth in capital includes not just cash, buildings, and machines, but also patents—*for the duration of the grant.* That's how a corporation reports it on a balance sheet, and that's therefore how it is accounted for in estimating national growth as well.

But all the patented knowledge that was originally counted as growth in capital becomes part of the residual once the patent expires, and what it loses in the value it had to its original inventor is gained by the inventor's nation. Just as public domain books, such as the Bible and the works of William Shakespeare, are both more numerous and more valuable than the universe of copyrighted ones, the universe of useful knowledge is a lot bigger than the universe of patented ideas.

How much bigger? Solow attempted to put a number on the rate of growth in formerly (and also never) patented knowledge—which included everything from calculus to the laws of motion—and assumed, for the sake of simplification, that the rate of increase was not only regular, but independent of changes in custom, law, or historical contingency; that is, knowledge, like Topsy, just "grow'd." Such simplifications are essential for theory building, but this particular one just pushed the big question back another step: Since knowledge, whether patented or not, is rarely lost, and the sum available has been increasing at least since the invention of written language more than five thousand years ago, what caused it, for the first time in history, to increase faster than the rate of population growth?

Population and prosperity *are* correlated, albeit imperfectly. Adam Smith was the first to recognize the hugely important but completely obvious correlation (this is a pretty good definition of genius) when he pointed out that the value of specialization utterly depends on the size of the community in which one lives. A family living alone grows its own wheat and bakes its own bread; it takes a village to support a baker, and a town to support a flour mill. Some critical mass of people was needed to provide enough customers to make it worthwhile to invest in ovens, or looms, or forges, and until population levels reached that critical level, overall growth was severely limited.

That level, however, was reached long before it had any impact on per capita growth in productivity. From 1700 to 1820, China grew in population from 138 million to 328 million, which increased its production of goods and services from $83 billion to $229 billion—but both population and production increased at almost exactly the same rate, about 0.75 percent annually. Population growth alone is clearly not sufficient to explain, for example, how the population of Britain, during the same period, could increase at the same rate as China's, but its gross domestic product nearly one-third faster. Something more than specialization through population growth was at work.

This was the conclusion of E. J. Hobsbawm, who argued that the fuel for the Industrial Revolution was not coal but demography: population—or, more precisely, the growing size of markets both domestic and foreign. While recognizing, generally, the importance of a cadre of mechanics to build and repair steam engines, forges, lathes, and spinning machines, he minimized the importance of the inventions on which they practiced their trade. The presence of a thousand brilliant inventors was far less important, in Hobsbawm's words, than "the mass of persons with intermediate skills, technical and administrative competence . . . without which any modern economy risks grinding into inefficiency."

Another currently fashionable explanation for the Industrial Revolution is also a demographic one, though subtle: not the nation's *overall* birth rate, but the birth rate in a particular subset. This is essentially the theory proposed in Gregory Clark's examination of the relationship between differential reproduction in different classes.

Clark's discovery, arrived at by combing through centuries' worth of parish records, was that the wealthiest families in England were far likelier to have more sons than similarly wealthy families in China, and therefore had less real property to pass along to the average member of the next generation. Throughout the period 1540–1640, preindustrial Britain exhibited a fair bit of both upward and downward mobility, largely due to primogeniture, which obliged a prosperous landowner to leave his land to only one of his sons—and the typical landowner had up to eight. Since all but one of those sons would have to find his own niche in the economy, "craftsman's sons became laborers, merchant's sons petty laborers, large landowner's sons smallholders" carrying with them the habits of hard work, deferred gratification, literacy, and a disposition to settle disputes peacefully, all of which showed a decided increase during the eighteenth century. As upper-class habits trickled throughout society, so did economic growth.

This theory explains fairly well why the son of a country squire might find himself learning the craft of a carpenter, and it may explain some critical aspects of historical growth in national wealth, particularly in Britain. A literate population bound by the rule of law and exhibiting middle-class behaviors such as deferral of gratification and low levels of corruption is as valuable to a nation as it is to a business firm. A recent World Bank analysis found that a significant amount of wealth worldwide is derived from such behaviors, and that human capital and the rule of law might account for up to 30 percent of the residual.

But the middle-class work habits that proved so vital for a new generation of factory laborers were valuable only because there were factories for them to work in. And Boulton and Arkwright weren't motivated to build the factories because good workers suddenly became available, but because they had machines that needed housing. Similarly, whatever percentage of Solow's residual is attributed to a reliable labor force operating in a relatively uncorrupt economy ruled by law, something still needs to explain why growth in prosperity, three-quarters of it derived from the creation of useful knowledge, stagnated for five millennia and then exploded in Britain in the eighteenth century.

This is where Solow's simplifying assumption—the idea that the growth in useful knowledge occurs at a constant rate, directly proportional to population—however analytically valuable, runs out of gas. It can't explain the steam engines and reciprocating chisels of the Portsmouth Block Mills, or the boring machines of Bersham Furnace, or the water frames in the Derwent Valley, because it implies that the decisions made to investigate the properties of steam, or iron, or silk were just as probable in eighteenth-century India as in eighteenth-century Europe; more likely, really, since India was home, in 1700, to more than thirty million more potential inventors than all of Europe, including Russia. The only thing that can be said in favor of making the growth of knowledge what economists call an *exogenous* variable (i.e. one without deliberate purpose, a kind of magical growth in knowledge from which everyone could benefit) was that its constant rate of increase made the equations simpler. By the 1980s, however, a number of economists examining the eighteenth century's wealth explosion thought that Solow's equations might be too simple.

The best known of them, Paul Romer of Stanford, made his

reputation by demonstrating that useful knowledge—the largest component in Solow's residual, and therefore the most important component of any increase in national wealth—doesn't accumulate by itself, independent of the larger economy, but rather is almost entirely dependent on decisions made by individuals seeking some sort of economic advantage. Romer showed that such growth depended on real people acting, in their perceived self-interest, to create knowledge. His model was both stable and reflective of the real experience of economic growth since 1800. It also resolved, in a mathematically rigorous manner, the conundrum in Solow's fundamental growth equation, which was precisely how the initial stage of knowledge creation benefits the investor/inventor and therefore is counted in the "capital growth" segment, while the larger sum spills over into the economy at large and forms the residual.

More important, Romer demonstrated that the growth in useful knowledge occurred at anything *but* a constant rate; that it was, instead, highly sensitive to the economy in which it occurred. Romer recognized that the creation of thc idca behind a new invention, or product, was just another fixed cost, like constructing a building, or buying a machine; it was just as expensive in skull sweat for James Watt to design the first Watt linkage as to manufacture a thousand. Because knowledge is the sort of property that can be sold to multiple consumers without lowering the value to any of them—Romer termed it *nonrivalrous,* as distinct from tangible, or *rivalrous,* property, which can be sold only once—the payoff for anyone producing it can be very large indeed.

At least, in a large enough population.

IN 1993, A DEVELOPMENT economist named Michael Kremer published a paper in the *Quarterly Journal of Economics* that examined population growth over time and tracked it against the expansion of what Romer called nonrivalrous, useful knowledge, and concluded that the creation of knowledge is directly proportional to the size of the population. Kremer's model made two assumptions: first, that inventive talent and motivation are randomly distributed throughout any population; and, second, that the larger the population, the larger the total output of inventions. The idea—that since each person has the same likelihood as any other of inventing the wheel, or the steam engine, or the iPod, more people means more inventors—may seem counterintuitive; it took only one genius, after all, to put stirrups on a saddle. But it works. A nation needs only twelve players to field a team to compete for the Olympic gold medal in basketball, and even tiny nations would find such a number easy enough to assemble. But only a few nations are actually competitive, because the twelve players represent the top of a pyramid of competition in which the larger the base, the greater the height.*

Thus, from the first century, when the world's population was around 128 million and its GDP about $105 billion,† to 1500, when population was 438 million and GDP $238 billion, the various growth equations demonstrate that all of the components of economic growth—land per worker, capital per worker, and therefore the residual—increased at a constant rate. More workers, more income, and, by inference, more invention. Moreover, the country-by-country numbers are relatively close in per capita terms; in 1500, each of the 48 million residents of Western Europe was producing, on average, about $772 annually, while China's 103 million were good for about $600 a head.

The year 1500 is significant because it's the one that marks what the economic historian Kenneth Pomeranz calls "the Great Divergence." Up until 1500, the difference between GDP per capita in the world's poorest countries and the richest was in general less than 50 percent; by 1820, it was more than 300 percent.

* In fact, the same process works in reverse, though at the expense of the pyramid metaphor. Not only does a larger population result in more inventions, the wealth created by those inventions permits even more population growth, and yet another Malthusian trap.

† In units called Geary-Khamis dollars, each of which is equivalent in purchasing power to a U.S. dollar in 1990, and which is the benchmark "currency" of choice for really long time ranges.

After 1500, the lockstep increases in population and knowledge ended. From 1500 to 1820, China's population nearly quadrupled, and so did its GDP, with no increase in per capita GDP. During the same period, the population of France doubled, but its per capita GDP increased 56 percent; that of Great Britain *quintupled*, and its per capita GDP increased nearly two and a half times.

	1500	1700	1820
POPULATION			
France	15M	21.47M	31.25M
Netherlands	0.95M	1.9M	2.33M
Britain	3.94M	8.56M	21.24M
China	103M	138M	381M
India	110M	165M	209M
GDP [in million International Geary-Khamis dollars]			
France	10,912	19,539	35,468
Netherlands	723	4,047	4,288
Britain	2,815	10,709	36,232
China	61,800	82,800	228,600
India	60,500	90,750	111,417
PER CAPITA GDP [in International Geary-Khamis dollars]			
France	727	910	1,135
Netherlands	761	2,130	1,838
Britain	714	1,250	1,706
China	600	600	600
India	550	550	533

Tables are all from Maddison, ed., *The World Economy: Historical Statistics.*

Kremer's theory was intended to explain productivity growth over a very long time frame; his article was entitled "Population Growth and Technological Change: One Million B.C. to 1990." However, for growth and change after 1500, and especially after 1700, it presents a pretty serious problem for the idea that inventiveness is directly proportional to population. Over the last five centuries, the theory only works by assuming that one can usefully calculate an average for the entire world; but, especially since 1820, the worldwide growth rate has been severely distorted by the huge acceleration in productivity in Europe and North America, which is a little like calculating the average wealth of the patrons in a restaurant before and after Bill Gates enters it.

It's not as if Kremer was unaware of the problem; he acknowledged that for centuries, poor but populous countries like China and India had experienced decidedly low research productivity, which at least suggested that inventiveness is a function of income rather than population, but not enough to change his basic thesis. That thesis, however, is considerably more persuasive when the time span is more than a million years than when one is examining a single century or a single country. But if a large population alone doesn't improve the chances of the steam engine's being invented in any particular nation, what does? Why did none of China's 138 million people invent a working steam engine, while Thomas Newcomen, one of fewer than nine million Britons, did?

China's size, its extraordinary technological head start, and its relative isolation have made it, for centuries, an irresistible laboratory for theorizing about industrialization and invention. Most of the resulting theories end up emphasizing either China's geography, science and technology, demography, or political culture.

Geography first. Geographical determinism is the notion that mineral resources, topography, and climate are the key drivers of human history. It remains eternally popular as an explanation for everything from political conflict to pandemics; but China's missing Industrial Revolution looks like a poor example of it. Pomeranz attributes a good deal of his "Great Divergence" to the relative ease with which British coal was extracted and therefore iron produced. But China had, and has, huge coal deposits, just as close to the surface, and China's forges were using coal to produce as much iron in the year 1080 as all of Europe did seven centuries later. They even did so for the same reason: the Yangtze delta had been deforested by the same demands as were the English Midlands, namely, construction, heating, and smelting. And even though the barbarian invasions of the twelfth and thirteenth centuries pushed China's center of gravity south to less coal-rich areas (China's nine southern provinces have less than 2 percent of contemporary China's coal reserves), iron production rebounded almost immediately and by 1700 was certainly larger in China than Europe. Neither Europe nor Britain truly had, to an objective eye, any advantages in mineral wealth, climate, or navigable waterways.

The state of Chinese science relative to the West seems a more promising explanation than its geography. It seems plausible, for example, to argue that if Chinese scientists failed to uncover the foundation principles unearthed by Torricelli and Boyle, Chinese inventors would have found it far more difficult to replicate the work of Newcomen and Watt. This is a small part of what the eccentric Cambridge don Joseph Needham spent his life investigating and chronicling in the two dozen volumes of his masterwork, *Science and Civilization in China*. Needham's conclusion, unfortunately, doesn't do much to confirm this particular diagnosis: he argues that the basic understanding of both vacuum and adiabatic pressure—the phenomenon that causes a gas to cool when it expands and heat when it is compressed—was present in China from the thirteenth century onward. Even if this overstates the case, a reasonable consensus exists for the belief that by the sixteenth century China had enough awareness of atmospheric pressure to produce not only a Newcomen-style reciprocating steam engine, but one that produced rotary motion.

Chinese engineers had already shown remarkable cleverness in transforming rotary into reciprocating motion. By 400 CE they had developed a system of water "levers," which used a waterwheel to fill a chute with water, tipping it first in one direction, then in another. Even more significant, and well known, was China's twist on the blast furnace, which, as the careful reader will recall, was being used to make cast iron in China by 200 CE, at least a thousand years before one appeared in Europe. The Chinese version of this device, called a box bellows, used a piston to pump air in and out of a cylindrical box. Early box bellows were hand-operated, but, in legend at least, sometime around 30 CE a Chinese inventor named Tu Shih hooked the piston up to a waterwheel-driven crank, thus facilitating the transfer between reciprocating and rotary motion seventeen centuries before Watt and Pickard. But it was an English historian named Ian Inkster, writing in 1842, who realized the potential for a box bellows piston that produced force on both strokes, writing, "let it [the box bellows] be furnished with a crank and flywheel to regulate the movements of its piston, and with apparatus to open and close its valves, then admit steam through its nozzle, and it becomes the double-acting engine." Pomeranz agreed: "The Chinese had already recognized the existence of atmospheric pressure, and had long since mastered (as part of their "box bellows") a double-acting piston/cylinder system much like Watt's, as well as a system for transforming rotary motion to linear motion that was as good as any known anywhere before the twentieth century. All that remained was to use the piston to turn the wheel rather than vice versa."

That was all that remained. It still remained when the first steam engines in China were being imported from Britain. The box bellows was undeniably an inventive leap, but not in any useful direction. The bellows, it turns out, is not a mirror image of the piston, but its opposite. To the degree that it works as a bellows, it cannot work to drive a piston. The Chinese could have a bellows, or a vacuum, but not both, at least not without a strong theory of the behavior of gas and pressure like the one articulated by Boyle in the seventeenth century.

And they didn't have one. Chinese science had taken a very different path than its analogue in the West; most especially it lacked the strong historical foundation—no Descartes, no Galileo, no Bacon—needed to develop the experimental science of the Enlightenment. Even worse: When a scientist like the polymath Fang Yizhy attempted to import Western methods into China in 1644, producing a mammoth collection of works on mathematics, engineering, and natural philosophy entitled *The Small Encyclopedia of the Principles of Things,* it lacked any readership outside the aristocracy; China's master artisans were so severely handicapped by illiteracy that eighteenth-century Chinese literacy rates never got much higher than 40 percent, at a time when 60 percent of British men were able to read.

If the Chinese handicap was neither geographic nor scientific, was it demographic? Even with Chinese literacy rates barely two-thirds that of Britain's, by 1700 there were still more *literate* Chinese than the entire populations, literate and otherwise, of France, Germany, Italy, and Britain combined. How is it possible that not one of them was capable of inventing a spinning jenny? The only difference between Hargreaves's invention and the machine used by Chinese cotton spinners, for example, was the draw bar, a device whose "fingers" could pull a large amount of softly wound cotton. The draw bar was not a complicated device, and yet even though at least five times as many Chinese as Britons were spinning fibers into yarn, there is no evidence that any of them invented one.

Instead, it appears that China's huge population, and its powerful central government, both of which would seem to be advantages for industrialization, were liabilities when combined. The population's demands for food were a tiger by the tail for the imperial court for centuries; so many mouths to feed meant that the investment capital available demanded single-minded focus on agricultural innovation. Just as contemporary scientists and inventors "follow the money," writing proposals in the subjects that granting agencies currently favor, so Chinese innovation pursued those avenues for which support was available. The difference is that contemporary scientists can select from many patrons; in China there was only one.

As with Tudor England, government monopoly of patronage meant control. Virtually all copies of the seventeenth-century Chinese encyclopedia, the *T'ien Kung K'ai-wu* or *Exploitation of the Works of Nature,* which included illustrations of everything from hydraulics to metallurgy, were destroyed because, according to Joseph Needham, much of the material touched on industries that had been granted monopoly status by the Qing emperors: "The absence of political competition did not mean that technological progress could not take place, but it did mean that one decision-maker [i.e. the Emperor] could deal it a mortal blow." It is therefore no surprise that a high percentage of both the inventions and inventors we associate with China from the time of the Han Dynasty to the Qings were government sponsored and employed.

Another liability of a strong central government is that it is, well, strong. Europe's fragmented system of sovereign states made it possible for innovative minds such as Paracelsus, Leibniz, Rousseau, and Voltaire to "shop" for more congenial places whenever they skated too close to heretical or otherwise challenging notions; in China, one had to travel a thousand miles to a place where the empire's writ ran not. And since China, perversely, was able to keep from plummeting down a Malthusian hole by using its enormous geographic extent to expand land under cultivation in the southern and western territories, technological stagnation, until contact with the West, seemed to have few if any costs.

In the end, however, neither territory, nor politics, nor even science is as powerful as culture in explaining China's inability to produce its own steam engines, puddling furnaces, or spinning jennys. Bertrand Russell translated the Chinese term *wu wei* (usually "doing without effort") as "production without possession"—a simplification, no doubt, but one with some powerful resonance to the notion that China's great burden was a lack of the itch to own one's own work.

ON THE OTHER HAND, that itch was powerfully felt in the seven United Provinces of the Netherlands from their de facto founding in 1581, when the onetime Spanish colonies abjured the rule of Philip II. By the beginning of the seventeenth century, the Netherlands was home to Europe's most cosmopolitan culture; the world's first stock market; a large and powerful merchant class, including Europe's largest bankers; access to millions of consumers through its enormous merchant fleet and colonies; the rule of law; and tolerance for every religious confession in the world, including the best established and assimilated Jewish community in the world. It had developed, out of deference to the

North Sea, a huge network of windmills, canals, and waterwheels. It even had, from 1683 to 1688, the prophet of the concept of intellectual property, John Locke, who wrote most of the *Treatise on Government* while living in Utrecht, Linden, and Amsterdam.

And it had a deep respect for inventions and inventors. Corneliszoon's crank-operated sawmill was far from an unusual case. From 1600 to 1650, the Dutch government issued between five and ten technical patents annually. However, the pace thereafter dropped off precipitously—and the reasons are telling. Petra Moser, now a professor at MIT's Sloan School of Management, spent four years examining more than 15,000 different inventions exhibited at nineteenth-century world's fairs, and their equivalents, and discovered a fact that seems at first glance to discredit the idea that patent protection was essential for innovation: Nations without patent laws were in many cases just as inventive as those with them. Or even more inventive; some of the nations best represented at those industrial fairs actively *discouraged* the patenting of inventions.

The reason seems to be that whether or not they enforced a patent law, smaller nations or domains, such as the Netherlands and Switzerland, were vulnerable to the theft of their innovations by competitors in larger nations. The bargain of patent protection runs two ways: The state, in return for making an idea public, offers legal recourse to its creator should someone within the state steal the idea. Since making one's invention public in a nation with patent protection offered protection against theft only up to its own borders, only a large nation offered a large enough market to make the deal a good one, and (in Moser's words) the small nations "would have been silly to patent [their] innovations."

This logic inhibited investment in entire categories of innovation. Those nations that relied on secrecy rather than patent tended to specialize in the sort of inventions that cannot be easily reverse-engineered, such as chemicals or dyes. One consequence is that almost all mechanical inventions—particularly, all steam engine innovations—were produced in countries that enforced some sort of patent law, since one can scarcely sell a compound engine and simultaneously keep its workings secret. Another is that any benefit from the cross-fertilization of ideas resulting from the public requirement of patent law, including publication of specifications, was lost.

The result is that while the Netherlands led the world in per capita (really, per man-hour) GDP every year from 1700 to 1820, its average compound growth rate over that period was actually negative, falling more than 13 percent. From there on, Britain took over the lead, with a growth rate per man-hour of 0.5 percent—a rate that exploded to growth of 1.4 percent annually from 1820 to 1890. The remarkable growth of the Netherlands during the 1600s essentially stopped a century later, and the only persuasive reason is size, or rather scale. A small country can shelter the world's largest banks, shipbuilders, and even textile manufacturers, but since it can protect inventors only from their own countrymen, growth that depends on the creation of new knowledge is fundamentally unsustainable, like a nuclear chain reaction with insufficient critical mass. Just as fission requires that a sufficient number of uranium nuclei be, in some sense, accessible to the others, a chain reaction of innovation sustains itself only if innovations are accessible to one another. A few thousand Europeans, no matter how inventive their work in chemicals, or metallurgy, could not create an Industrial Revolution unless they could inspire (or borrow, or even steal) from one another; a few thousand Britons, precisely because they concentrated their efforts on "public" inventions, most especially the steam engine, emphatically could. The consequence is that smaller nations, by avoiding large-scale mechanical invention out of fear that their own territory was too small to make them profitable, deferred industrial leadership to those large enough to take the risk.

So if the Netherlands was too small, and China too big,* what nation was, in the immortal words of Goldilocks, "just right"? The most intriguing candidate would seem to be France, which had

* Perversely enough, nineteenth-century Russia was both too big *and* too small. In 1766, the brilliant Ivan Polzunov, inventor of the world's first two-cylinder engine, died of tuberculosis, and when his engine, which operated the bellows in Russia's largest forge, needed repair three months later, no one could be found who understood how to reassemble it.

just about every advantage that Britain had, and a lot more Frenchmen to exploit them. France's economy grew at almost precisely the same rate as Britain's between 1700 and 1780, and from a far larger base; only about a tenth of one percent per year less growth in industrial (and, for that matter, agricultural) output in France during those eight decades. In 1789, the year of the Revolution, France's foreign trade was 25 percent *greater* than Britain's, and its population was nearly three times bigger.

By the same year, however, Britain had a significant lead in any number of significant indices: a third more GDP per capita, far higher rates of urbanization (nineteen of every hundred Britons lived in cities, and only eight Frenchmen), nearly eight times as much money in the hands of banks, and a tax rate less than a third of France's. Thus, in part because of lower interest rates from the middle of the eighteenth century forward, the *availability* of capital (as opposed to its absolute amount) was significantly greater in Britain.

But it was not only, or even mostly, an advantage in business sophistication that gave Britain a head start on industrialization. Nor was it scientific sophistication, a yardstick on which France was *way* ahead: Watt was simultaneously a brilliant engineer and a gifted scientist, but he still needed to study French engineering texts because there were so few English ones available. His experience remained true through most of the nineteenth century, when French, and later German, scholars were giving a scientific foundation to the laws of thermodynamics and kinematics.

And they were doing so in an environment in which standardization was very highly valued, and strictly enforced. Possibly because it achieved status as a coherent nation-state centuries before England, to say nothing of Britain, France has a far longer history of activism in setting national standards; the Académie Française, as a case in point, has been protecting the purity of the French language since 1634. The project of synthesizing the knowledge held tightly in the hands of French artisans, mechanics, and craftsmen began only forty years later, when the Académie des Sciences—France's equivalent of the Royal Society—began a national "Description des arts et métiers" intended to establish standard versions of hundreds of apparatuses. Among other things, the project provided Diderot and his *Encyclopédie* with more than 150 drawings and engravings of water pumps, looms, and forges. In the following century, the French government explicitly took on the responsibility of educating and training engineers with the founding of several schools focused on applied science, including the École des Ponts et chaussées in 1774; the École polytechnique was founded by a graduate of Ponts twenty years later, in part to impose technical standards on industry with the same rigor that the Académie governed the French language. The *Encyclopédie* itself promised "to offer craftsmen the chance to learn from philosophers, and thereby hopefully to advance further toward perfection."

In Great Britain, on the other hand, inventions were much more of a haphazard process, performed by onetime wheelwrights and carpenters competing, rather than collaborating, with one another. Their success did not go unnoticed in France, nor unremarked. In 1824, an École polytechnique graduate named Sadi Carnot wrote *Réflexions sur la puissance motrice du feu et sur les machines propres à développer cette puissance*—the first theoretical explanation of the thermodynamics of steam power—essentially out of dismay that the great achievements of British engineering had been produced by men, like Watt, with no formal schooling. The snobbery served French science well; less well French innovation. If one secret to sustaining an inventive culture was making inventors into national heroes, it was a secret that didn't translate well into French. Between 1740 and 1780, the French inclination to reward inventors not by enforcing a natural right but by the grant of pensions and prizes resulted in the award of nearly 7 million livres—approximately $600 million today*—to inventors of largely forgot-

* It is even harder to calculate the current value of eighteenth-century French currency than British. From 1726 on, an ounce of gold was set equal to 92.5 livres; during the same period, an ounce of gold cost £4.10. Seven million livres, therefore, equaled about £310,000. Using the index of average earnings, this is more than £400 million today.

ten devices, but Claude-François Jouffroy d'Abbans (inventor of one of the first working steamboats), Barthélemy Thimonnier (creator of the first sewing machine), and Aimé Argand (a partner of Boulton and friend of Watt whose oil lamp became the world's standard) all died penniless. Other than Joseph-Marie Jacquard, the creator of the eponymous loom, and perhaps the Montgolfiers, the French did not lionize their inventors.

This didn't mean they didn't understand the strategic importance of technology. Carnot himself wrote, "to deprive England of her steam engines, you would deprive her of both coal and iron; you would cut off the sources of all her wealth, totally destroy her means of prosperity, and reduce this nation of huge power to insignificance. The destruction of her navy, which she regards as the main source of her strength, would probably be less disastrous." Competition with Britain alone might have allowed French industrialization to survive the haughtiness that made the nation elevate pure science over its commercial applications, if not for an unfortunate bit of timing. The same year that Joseph Bramah was hiring Henry Maudslay to help build his locks at 124 Piccadilly, several thousand citizens of Paris marched down the Rue Saint-Antoine to the nearly empty prison known as the Bastille. The same year that the British government was certifying a grant of incorporation for Boulton & Watt to design and sell steam engines, the French government was beheading Antoine Lavoisier, the chemist whose research on heat was central to the theory behind those engines.

It was not immediately apparent that the French Revolution would be hostile to invention or inventors. The first law protecting intellectual property in France was passed in 1791, in ringing language that declared,

> every novel idea whose realization or development can become useful to society belongs primarily to him who conceived it, and that it would be a violation of the rights of man in their very essence if an industrial invention were not regarded as the property of its creator.

Unfortunately for the cause of innovation, the law was abrogated only two years later, as a side effect of the extreme violence of the Terror. And while it was reinstated in 1794, nobody seems to have told the French patent office. They were already playing catch-up in 1792, when Britain granted 85 patents, and France, with a population twice as large, issued 29. In 1793, that number fell to 4. From 1793 to 1800, in fact, Britain issued 533 patents to France's 65. In addition to all their other world-historical effects, the French revolutionists, and the Corsican emperor whose wars were the Revolution's last chapter, constitute the most important reasons that Britain and America established a thirty-year lead on all other European nations in the development of steam power. As Jeff Horn, whose study of French industrialization is very close to the last word on the subject, put it, "When the revolutionary and Napoleonic wars ended in 1815, the British were approximately a generation ahead in industrial technology and in the elaboration of the mechanized factory."

Looked at through the history of ideas, the French attitude toward invention, and even its revolutionary spirit, share a common origin. To the same degree that Britain's beliefs about property are traceable to John Locke's *Second Treatise on Government,* France's can be found in the *Discourses* of the onetime engraver, writer, musician, and philosopher Jean-Jacques Rousseau.* In his *First Discourse,* for example, Rousseau shared his discomfort with technical progress, which he associated with decadence and moral decline; his *Second Discourse* argued that the invention of *any* technology, by demonstrating that some are more gifted than others, promotes inequality and eventually tyranny: "Astronomy was born from superstition . . . physics from vain curiosity" (*First Discourse,* volume I). In his "treatise" on education, *Émile,* Rousseau attacked what has come to be known as amour propre: the invidious striving after excellence in the eyes of one's fellows.

* The Locke versus Rousseau debate remains one of academe's almost preternaturally popular, used to explain everything from the differences between modern and medieval perspectives to the reason for Britain's reluctance to adopt the euro.

Rousseau's fetish for compassion and equality have made him a powerful influence on generations of Marxists, but his earliest, and most consequential, impact was on the revolutionaries of eighteenth-century France. When in 1793 the Jacobins closed the Académie des Sciences on the logic that "the Republic does not need savants," they were channeling Rousseau.

And they were hampering their Republic in the race to technological mastery. The finish line for the first stage of that race had been the use of condensed steam to convert atmospheric pressure into the reciprocating motion of Newcomen's pumps. For the second stage, it was converting the expansive power of steam into rotary motion able to drive dozens, and then hundreds, of spinning and weaving machines.

The third stage was converting steam power into motion. Locomotion.

CHAPTER TWELVE

STRONG STEAM

concerning a Cornish Giant, and a trip up Camborne Hill; the triangular relationship between power, weight, and pressure; George Washington's flour mill and the dredging of the Schuylkill River; the long trip from Cornwall to Peru; and the most important railroad race in history

THE CAUSE OF THE ACCIDENT at Poldory Mine in January 1784 is not precisely known, nor is its date; some accounts have it occurring on the sixteenth, others on the nineteenth. Its effect on the mining families of Cornwall's Gwennap Valley, about two miles from the town of Redruth, was depressingly routine: tragic for the six families that lost their breadwinners, a reminder to everyone else of the dangers of mining—cave-ins, floods, fires, and suffocation.

The risk of each hazard increased in direct proportion to the mine's depth, and by 1784, Cornwall's copper was being carved from seams several hundred feet below the surface. This made Poldory, and its neighbor, the Ale-and-Cakes,* utterly dependent on the Boulton & Watt engine pumping the water out of the two

* The names of eighteenth-century Cornish mines are as personal, and as obscure, as the names given to thoroughbred racehorses and recreational sailboats.

mines and into the Great County Adit that drained a good portion of central Cornwall. If the engine didn't work, neither did the miners.

The reliance on steam power, however, had introduced a new danger: catastrophic failure of the engine itself. Not all steam engine failures are catastrophic—complete and sudden. Valves can stick, or beams crack, harming only the engine itself. A failure of the engine's cylinder housing or boiler is different. The sudden release of steam under pressure is literally explosive, sending shards of metal flying outward at several hundred feet per second, driven by jets of scalding steam. It was precisely this sort of accident that killed three miners, and maimed several more, at Poldory, and when the news reached the engine's designer and builder, his reaction was predictable. On January 24, James Watt wrote to Thomas Wilson, a mine owner living in the town of Truro in Cornwall,

> I am exceedingly shocked at the account of the accident at Poldory and should have been Glad to have had some particulars. They must certainly have had a very strong steam otherwise, the people would have had time to escape. Please also to advise who the people were and how so many came to be about the boiler; Copper tubes must be entirely given up without men can be found more carefull [*sic*] in the management of them. If any of the families of the deceased or the surviving persons who were scalded are in distressed circumstances, I am sure that Mr. B[oulton] will Join me in being pleased that you should give a small matter for their immediate relief as if of your own accord without mentioning our names . . ."

The phrase "strong steam" is telling. By 1784, Watt was decades removed from his first experience with Newcomen engines, but those decades had done nothing to ease his fear of the caged power of steam under pressure. Some of the concern dates to his earliest experiments: two years of sealing materials that failed to seal, tubing that leaked, and cylinders with seams that burst. The larger part, however, was an analytical blind spot, one that he shared with the most sophisticated scientists of the eighteenth century.

That blind spot, about the nature of the relationship between heat and motion, was no longer the belief that phlogiston was released whenever anything was burned. The failure of phlogiston theory to account for any number of observed phenomena had opened the door for another, more useful though still flawed, to take its place. This was the idea that heat remained a substance, but a weightless one, called *caloric.* The Scottish philosopher William Cleghorn, another protégé of Joseph Black, theorized that this "subtle fluid" (in the words of Lavoisier, who also coined the word "caloric") was a gas whose properties included varying levels of attraction to different types of matter, thus explaining why the heat capacity of coal was different from that of glass. The theory further held that caloric could be neither created nor destroyed, but only changed from Black's latent heat—the potential locked up in a combustible substance—to sensible heat, and then back again, with the total amount of caloric in the universe staying constant.

Caloric theory remained the conventional wisdom for sixty years or more for a simple reason: It explained, better than any alternative, dozens of physical phenomena. Hot fluids cool, in caloric theory, because caloric repels itself, thus diffusing from an area of high concentration to a lower one. It explains heat radiation and Boyle's Law, and even formed the basis, forty years after Poldory, for Sadi Carnot's *Réflexions,* and the first working theory of steam engines: that their capacity depended only on the difference between high temperature and low temperature, which, in Watt's steam engine, was the difference between the temperature of the boiler and that of the condenser.

This didn't mean that no one was thinking outside the caloric box. There was, for example, the thoroughly remarkable Benjamin Thompson of Massachusetts, a loyalist American who, after backing the losing side in the Revolutionary War, moved, first to England (where in 1779 he was made a Fellow of the Royal Society), and four years later, apparently on a whim, to Bavaria. There he found himself, on behalf of the Prince-Elector Karl-Theodor, running an espionage network that stole design sketches from the Soho Manufactory and spirited them out of England. For this and other services (including the invention of Rumford Soup, a concoction of peas, barley, potato, and old beer intended to meet the nutritional needs of Europe's poor) he was made a Count of the Holy Roman Empire in 1791.

Seven years later, Count Rumford (as he had styled himself, after the New Hampshire town where he had been a schoolmaster) performed an experiment investigating the nature of heat, inspired by his observation of Bavarian metalworkers boring out a brass cannon barrel, which generated enough heat from friction that the barrel was too hot to touch. Lavoisier's theory predicted that the caloric associated with drilling should have melted the brass shavings into which it presumably had been transferred. Since it did not, Rumford tested it, rather cleverly, by boring a cannon barrel underwater. The borer's friction produced enough heat to keep the water boiling—and the water continued boiling as long as the borer was spinning, thus disproving the idea that caloric was a property contained within matter, since it was never exhausted.

If caloric was not a fluid, it had to be something else. Rumford's monograph, *An Experimental Enquiry Concerning the Source of the Heat Which Is Excited by Friction,* argued that the "something else" was actually motion: that heat and motion are essentially the same thing. This was critical, and surprisingly slow in coming. It is, after all, not hard to find places outside the laboratory where mechanical work creates heat: rubbing two Boy Scout–approved sticks together, for example. John Locke himself observed the heat produced by the mechanical energy of a wheel rubbing against its axle.

But even though Count Rumford shot a very large hole in the idea of caloric, theories are overturned only by better theories, and he didn't really have one to offer. So long as caloric theory was still an effective way of approximating physical reality, it wasn't going away. This was a real obstacle to engineers, since a key element of caloric theory confused the nature of work in steam engines. Caloric theory held that heat was latent within combustible materials and was simply converted from latent to sensible, which "proved" that there could be no advantage to a high-pressure engine, which would simply increase sensible heat at the expense of latent heat.

This was confounded by another blind spot: Until the middle of the nineteenth century, scientists and engineers alike widely believed that the source of a steam engine's work was the pressure of expanding steam on a moving piston. More pressure, therefore, equaled more work. So far, so good. But the analogy they used as a model was waterpower, and that was not so good. A waterwheel produces the same amount of work when relatively little water flows over a long distance as it does when a lot of water flows over a short distance; a waterwheel will make as many revolutions when it catches a trickle that has fallen a thousand feet as when it is driven by a river falling a few inches. By the same token, they reasoned, a steam engine would be just as efficient using low pressure to drive a long beam as high pressure to drive a shorter one—and steam, as the families of the Poldory miners would certainly testify, is a lot safer at low pressure.

Reasoning by analogy is always suspect. What was missing in this example was an understanding that it was not the *pressure* that mattered, but the *heat*—heat energy and mechanical energy were just different forms of the same thing. The heat needed to raise one pound of water a single degree is equivalent to lifting 772 pounds one foot, and vice versa: lifting 772 pounds one foot generates an equivalent amount of heat.* The slow realization of this relationship is explained largely by the low efficiency of the early engines. If you're converting only 2 percent of the heat energy into work, you'd be hard pressed to notice its absence. And even when they noticed it, the blind spot endured. No matter how many times engineers observed more work being done with more heat, they were unable to make any sense of the results.

* Or, indeed, any form of thermal or electromagnetic energy. This particular bit of equivalence, the British Thermal Unit, is an early nineteenth-century measurement that has been mostly replaced by a frighteningly large array of units, including calories (and kilocalories), joules (and kilojoules), electron volts, kilowatt-hours, and therms, each of which can be converted to the others.

This blinkered view didn't change the need for steam engines to deliver more power, of course; water needed to be pumped from ever deeper mine shafts, factories needed more power to drive larger wheels. The response of Boulton & Watt was Archimedean: longer levers driven by ever larger condensing engines, using cylinders up to five feet in diameter, with strokes of nine or even ten feet. That these behemoths offered a pretty unattractive power-to-weight ratio didn't seem at the time to be much of a problem; low pressure—about 8 to 10 psi—was reliable, safe, and affordable. A multi-ton condensing steam engine could pull a bucket from a mine shaft five hundred feet deep, or run a dozen shafts at a cotton spinning factory. But it couldn't pull itself any distance at all.

It wasn't that no one thought about the possibility of steam locomotion. In 1784, only four months after Poldory, Watt himself described, in the same patent that included the parallel linkage, a piston-driven steam carriage that used "the elastic force of steam to give motion." In 1785, Boulton & Watt's brilliant engineer William Murdock actually built a "steam carriage" and for the first time filed a patent in his own name. He even built a scale model of his carriage—a cylinder with a diameter of three-quarters of an inch and a stroke of an inch and a half—and decided to exhibit it in London, though there is no reason to believe that it would have been a success. Asking a steam engine to move itself—to say nothing of cargo or passengers—meant making it powerful *and* lightweight; and the only way to do that was increasing the pressure in the cylinder. With his fortune tied to the future of low-pressure steam power, Matthew Boulton himself intercepted Murdock on the way to London and persuaded him to return to building stationary Boulton & Watt engines in Cornwall.

Which was where, in any case, the next revolution in steam engines was going to occur. There, and in America.

BY THE MIDDLE OF August 1787, the fifty-five delegates to the world's first, and most consequential, Constitutional Convention had been meeting in Philadelphia for three months of an extremely hot summer. They had debated judicial appointments, executive departments, and every conceivable duty of a national legislature. They had established the line of command for the new nation's armed forces and proposed a national postal service. Given all that, it's no great shock that the first acknowledgment that the newly proposed federal government had any place in protecting the activities of inventors didn't come until August 18, when James Madison of Virginia proposed that the national legislature be empowered to "encourage knowledge and discoveries." The same day, Charles Pinckney of South Carolina submitted a proposal that the government be able "to grant patents for useful inventions."

Four days later, on the twenty-second, the convention adjourned for the afternoon and headed to the banks of the Delaware River to see a demonstration of the power of such useful invention: a forty-five-foot-long boat that resembled an Iroquois war canoe, with six oars on either side. The motive force for those oars, however, was not muscle. It was steam.

The steamboat's inventor, a onetime clockmaker and silversmith from Connecticut named John Fitch, had turned a traditional twelve-inch condensing cylinder on its side and used it to drive a piston with a three-foot stroke tied to an eighteen-inch axle. In his own words, "Each revolution of the axle tree moves twelve oars five and a half feet. As six oars come out of the water six more enter the water; which makes a stroke of about eleven feet each revolution. The oars work perpendicularly and make a stroke similar to the paddle of a canoe."

Fitch's steamboat was not, as many histories have it, the world's first. In 1772, two ex-artillery officers in the French army, the Comte d'Auxiron and Charles Monnin de Follenai, received a fifteen-year exclusive license to run a steamboat along the Seine. Unfortunately, their first attempt, a marriage of a Newcomen engine to a Seine bâteau, was less than successful: the engine was so heavy it sank the boat. Slightly more successfully, in 1783, the Marquis de Jouffroy d'Abbans took a 140-foot boat mounting a Newcomen-style engine out on the Saône from Lyon. He did make it all the way back to the dock, where cheering crowds met it—just in time, before the engine's vibrations destroyed the boat.*

The great importance of Fitch's steamboat was not that it survived its inaugural trip; it was his audience, who were properly impressed with what turned out to be the steamboat's maiden voyage. One delegate, William Samuel Johnson of Connecticut, wrote on the twenty-third, "the exhibition yesterday gave the gentlemen present much satisfaction and will always be happy to give him every countenance and encouragement in their power which his ingenuity and industry entitles him to." Two weeks later, the Brearly Committee (named for its chairman, David Brearly of New Jersey, and also, and unfortunately, known as the "Committee of Leftovers") reported on fourteen proposals to the convention; the last one was a recommendation to "provide limited patents to promote science and arts." The patent clause was incorporated, without a single dissenting vote, into Article I, Section 8, paragraph 8 of the United States Constitution.

It's easy to see why the American revolutionaries were so taken with the British attitude toward intellectual property. In almost every relevant way, they *were* British. The common law was as well known on the banks of the Potomac as along the Thames. Virginians signed on to seven-year apprenticeships as carpenters, millwrights, and glaziers exactly as their counterparts did in Yorkshire (sometimes an apprenticeship would begin in the latter and conclude in the former). The éminence grise of the American Revolution, Benjamin Franklin, was not only a Fellow of the Royal Society but also one of the most prolific inventors of the entire eighteenth century; Thomas Jefferson, the revolution's intellectual soul, took enough time off from his writing and architecture to design revolving bookstands, copying machines, revolving chairs, and even a new and improved moldboard plow. If those models weren't sufficient, eighteenth-century America showed even more enthusiasm

* Some histories still insist that in 1543, a naval officer in the service of Charles V of Spain named Blasco da Garay used steam to propel a boat across Barcelona harbor, though the story has been thoroughly debunked for more than a century.

than Britain itself for the intellectual forebears of patent law. Locke was considerably more influential among the American constitutionalists than he ever was to English parliamentarians, and had even drafted the first constitution for the Carolina colony. And not just Locke: the *Mayflower* had carried a set of Coke's writings from the Old World to the New, and both Jefferson and Madison had gotten their legal training from reading them. In the eighty-fourth of the Federalist Papers, Alexander Hamilton compared Coke's 1628 Petition of Right to the Magna Carta.

Fitch's influence on the future of steamboats was less enduring.* He gave up on oars fairly quickly and experimented with circular paddles and even an early propeller screw, which he used for ferry service between Philadelphia and Burlington, New Jersey, but his failures outnumbered his successes many times over. Though he pursued a national patent, and de facto monopoly, on steam-powered water travel with monomaniacal zeal for years (much of his time exhausted in disputes over priority with another inventor, James Rumsey, who used a jet of water propelled by a steam-driven pump to drive a boat on the Potomac in 1787), the final award of patent, almost three years to the day after his demonstration for the delegates in Philadelphia, was so limited in its language as to be commercially useless.

The fact that a national patent was available at all, however, was the significant thing. One reason that those constitutionalists were even in Philadelphia that summer was the realization that the union originally established in 1777 under the Articles of Confederation and Perpetual Union was responsible for governing a nation that covered a territory bigger than France, Britain, Germany, and Spain combined, with fewer tools than a New England town meeting. The confederation—unable to tax its citizenry or

* In the east end of the North Corridor on the first floor of the Senate wing of the U.S. Capitol is a series of frescoes painted by the Italian émigré artist Constantino Brumidi, thematically coordinated with the specific duties of Senate committees; over the doors leading to Room S-116, where the Committee on Patents originally met, are three portraits. Two of the subjects—Benjamin Franklin and Robert Fulton—are as well known as any names in American history. The third is John Fitch.

even levy soldiers in wartime, powers that were reserved to the individual states—wasn't deriving much benefit from its size.

And while national defense was obviously a more urgent deficiency, economic issues were a close second, including the recognition that some national authority needed to promote, and protect, invention. In 1800, there were only a few more Americans living in the New World than Dutch living in the old, and we've already seen that the Netherlands, despite its great wealth, was still too small to support British-style inventing; Massachusetts, or Virginia, wouldn't stand a chance. Inadvertently, the constitution had stumbled on the fundamental issue of scale in intellectual property. The value of a bar of gold or a bushel of wheat—Romer's rivalrous property—is no greater in a large country than a small one. On the other hand, the value of a nonrivalrous patent or copyright increases in direct proportion to the number of people one can sue to prevent its theft.

Not everyone agreed, even in eighteenth-century America. Thomas Jefferson, most notably, was reflexively offended by even the slightest odor of monopoly; in a much-quoted letter sent to his friend Isaac McPherson, Jefferson wrote:

> If nature has made any one thing less susceptible than all others of exclusive property, it is the action of the thinking power called an idea, which an individual may exclusively possess as long as he keeps it to himself; but the moment it is divulged, it forces itself into the possession of everyone, and the receiver cannot dispossess himself of it. Its peculiar character, too, is that no one possesses the less, because every other possesses the whole of it. He who receives an idea from me, receives instruction himself without lessening mine; as he who lights his taper at mine, receives light without darkening me. . . . Inventions then cannot, in nature, be a subject of property.

Nonrivalrous property indeed; small wonder that Jefferson is regarded as the intellectual godfather of the twenty-first century's "information wants to be free" movement. Partly because of his resistance, the First Congress, which opened for business in March 1789, failed to consider a system for granting patents for nearly a year. The first American patent statute was, however, eventually passed, and was signed by George Washington on April 10, becoming the law of the land—or of twelve-thirteenths of the land, since Rhode Island hadn't yet ratified the Constitution.

The original procedure was fairly straightforward. Each patent application was sent to a committee of the United States Senate, who then referred it to the attorney general, who passed it to the president, who signed it and returned it to the secretary of state (in 1790, Thomas Jefferson). That was changed soon enough to the newly named Commissioners for the Promotion of Useful Arts: Jefferson, Secretary of War Henry Knox, and Attorney General Edmund Randolph.

The American system was simplicity itself compared to the contemporaneous British system, which was a thing of cartoonish complexity:

Step 1: Inventor prepares petition to the Crown, including an affidavit sworn before a "Master of Chancery"

Step 2: Master sends petition plus accompanying affidavit to the Home Office, who reads, endorses, and sends to the Attorney General and Solicitor General

Step 3: The Attorney General and Solicitor General review and, if they approve, return both petition and report to the Home Office

Step 4: Home Office prepares a warrant; sends to the King

Step 5: King signs, and Secretary of State countersigns, the warrant

Step 6: Secretary of State sends warrant back to Attorney General and Solicitor General; they prepare a bill describing the invention, and transcribe it onto the actual letters patent (written on parchment, in order to have the force of law)

Step 7: Attorney General and Solicitor General send parchment bill back to Secretary of State; King and Secretary again sign and countersign, thus making it, literally, a "King's Bill"

Step 8: Secretary of State sends the King's Bill to the Signet Office, which prepares an identical version, known as the "Signet Bill" (on parchment, of course; see Step 6)

Step 9: Signet Office sends Signet Bill to the Lord Privy Seal, who prepares a Writ of Privy Seal—yes, on parchment—and sends it to the Lord Chancellor with Signet Bill and Letters Patent

Step 10: Lord Chancellor "engrosses" Letters Patent on parchment with language identical to the Writ, dates it, and finally seals it

The system was not only absurdly complicated but outrageously expensive. In 1792, the official cost of a patent was £70 for England and Wales, but "gratuities" to every secretary, official, and even doorman standing along the way typically ran another £20; the tariff including Scotland and Ireland could easily exceed £300. The cost of a U.S. patent application, by comparison, was fifty cents; if the patent was awarded, the recipient owed the federal government two dollars, plus another dollar for affixing the Great Seal of the United States.

Even so, the American system was a little slow getting started. The United States issued only three patents during all of 1790. The first went to Samuel Hopkins for his method of making potash; the second, "for manufacturing candles," was granted to a Boston chandler named Joseph Stacey Sampson. Both patents ended up generating far more wealth as rare documents, sold and resold to collectors avid for the signatures of Jefferson and Washington, than they ever did for their inventors. By number three, however, the system had identified an inventor who would do as much as any man alive to put steam on the move: a Philadelphian named Oliver Evans.

EVANS WAS BORN IN 1755 in the colony of Delaware and was apprenticed at the age of sixteen to a wheelwright in Pennsylvania.

By the end of the customary seven years of training and toil, he had already built his first invention, a machine tool for making the leather "cards" used for removing unwanted material from wool and cotton. One of Evans's most distinctive skills was the ability to see machines as a series of related steps, and his tool was an early, but telling, case in point: an assembly line in a box, it first bent the wire, then cut it to the proper length, and finally punched holes in precut pieces of leather, on which it mounted up to one thousand wire teeth per minute, with all the steps driven by a single rotating wheel. The hand-cranked card-making machine, for which Evans sought and received patent protection in Delaware, was an immediate success; so successful, in fact, that copies of it turned up, within months—and royalty-free—in Massachusetts, an early reminder to the inventor of the problematic character of intellectual property.

Evans's next invention took the process of using one mechanical motion to drive a succeeding one a giant step further. Starting in 1788, he designed and built an automated mill that turned wheat into flour in a single continuous operation, illustrating, on a much larger scale, his gift for sequential mechanics. This time an elevator—actually wooden cups affixed to belts—lifted the unmilled grains of wheat onto a conveyor belt, which, in turn, was driven by a rod with lands cut into it: effectively a horizontal screw. The belt then pushed the wheat through the millstones into a hopper where a mechanical rake alternately stirred and sifted the flour. And once again a single rotating shaft synchronized all the operations.

Of all the components of the mill, only the hopper was truly original; but the real novelty, and value, of the invention came from the way it coordinated several existing mechanisms. Evans had once again exhibited a gift for seeing the big picture, literally, if uncomfortably: "I have in my bed viewed the whole operation with much mental anxiety." In the event, it was the "whole operation" rather than any specific machine that earned Evans his first patents: first a fourteen-year grant in Maryland and Pennsylvania, a little later one of the same duration in Delaware, and one in New Hampshire to run for seven years.

His receipt of U.S. patent number 3 in 1790 generated prominent clients, and equally prominent critics. Though George Washington installed an Evans machine at his Dogue Run mill in 1791, his fellow Virginian Thomas Jefferson was nonplussed by Evans's claim of novelty: "If the bringing together under the same roof various useful things before known entitled him [Evans] to an exclusive use of all these . . . every utensil of life might be taken for use by a patent. . . . I can conceive how a machine may improve the manufacturing of flour, but not how a *principle* abstracted from any machine can do it."

Whatever Jefferson's concerns about patents in general, and Evans's in particular, he discharged his responsibility for them with diligence, and eventually—once he conceded the success of the law in promoting invention—enthusiasm. Evans himself was certainly spurred by the financial incentives of patent ownership, though their promise, in his case, frequently overran their actuality. His goal had been to earn his living selling licenses to other millers, but he found them resistant to automation, and in a classic pattern he spent more in legal fees suing patent infringers than he earned in licenses, despite renewing his patents nationally in 1808. By then, however, though his grist mills typically ran on waterpower, he was fully entangled in the next stage of the steam engine revolution.

In his own later recollection, Evans's interest in using steam power for transportation dated back to his apprenticeship; in 1773, he had noticed a blacksmith's apprentice using steam as the propellant in a gun. By 1783, he was already attempting to patent a steam locomotive, though his application to the Pennsylvania legislature was rejected for lack of a working model.

Using steam to propel vehicles over land was considerably more difficult than doing so on water, for reasons of simple physics. A wheeled vehicle needs to combat both gravity and inertia, plus any number of sources of friction, particularly the wheel against the axle rod. By contrast, relatively little power is needed to propel a lot of weight on water, primarily because of the lack of any real changes in elevation, close to perfectly even terrain, and negligible friction; this is why shipping large amounts of freight has always been cheaper by water than by land. As a result, a separate condensing engine could drive John Fitch's steamboat. With the water supporting the engine's weight, enough power could be produced to drive the craft, though slowly: four miles an hour, a brisk walking pace.

Locomotive engines, however, needed to put out more power with less weight; to be, in short, efficient. One of Evans's great Usherian insights was that if one could dramatically increase the temperature of the steam, and therefore its pressure, the separate condenser could be dispensed with, at a huge weight savings. Another was that even though the steam, without a condenser, would simply disappear into the air once it had driven the piston, it would still use less fuel than was used up constantly reheating and recondensing. All he had to do was discover a way to increase the heat and pressure in the cylinder by an order of magnitude—but without turning it into a steam-powered hand grenade.

The key was the boiler. Low-pressure engines, from Newcomen to Watt, had boiled water in chambers whose basic design wasn't much different from that of an oversized teakettle: a pot, or "haycock" shape, in which the furnace was placed below the water. Despite a number of improvements, it had remained essentially unchanged ever since.

Evans's stroke of brilliance was to place his furnace *inside* a water-filled chamber: more surface area for the heated gas, more heat transfer to the liquid. With nowhere to go, the boiling water turned to steam whose pressure, now about fifty pounds per square inch, drove the piston, first in one direction, then in another.

In 1804, Evans applied for and received a federal patent for a high pressure "vibrating steam engine" that incorporated a "Boiler with the furnace in the center of the water enclosed in brick work." To Evans, like every other inventor of the steam age, heat meant fuel, so increasing the first meant adding more of the second. But he also realized "the more the steam is confined . . . the greater will be the power obtained by the fuel. For every addition of 30 degrees of heat to the water doubles the power [and] doubling the heat of the water increases the power 100 times. Thus, *the power of my engine rises in geometrical proportion, while the consumption of fuel has only an arithmetical ratio*" (emphasis added). Evans had discovered how to produce the same power from a 500-pounder that Watt was getting from a two-ton monster; and as a bonus, it burned less fuel for every increase in horsepower—a huge advantage in an engine that needed not only to be mobile but also to carry its fuel with it. It was, by any definition, an act of genius.

It was not, however, a success. The value of high-pressure steam was greatest where its attractive power-to-weight ratio mattered most—steam locomotion—and America was still several decades away from railroad building. Evans was, like most inventors, forced to invest his time where he could find customers. In 1805, he built a thirty-foot-long fifteen-ton steam dredge for the Philadelphia Board of Health, which was the first steam land vehicle built in the United States; the same year, he created the first significant improvement over Watt's linkage, using an isosceles triangle to connect two bars at a single pivot point, a design later known as the Russell linkage.*

However, also in 1805, he was the litigant in no fewer than four different suits, brought on both philosophical and technical grounds. One result was that his income was slashed, leaving Evans, in his own words, "in poverty at the age of 50, with a large family of children and an amiable wife to support." Despite his many triumphs, Evans's life seemed, to him at least, a succession of unsatisfying successes and paranoia-reinforcing failures. In 1795, he had published *The Young Mill-wright and Miller's Guide,* which was effectively an entire millwright's education between hard covers, went through dozens of printings, and was probably the bestselling "professional" book in the young repub-

* The Russell (or Scott Russell) linkage was actually invented and patented in 1803 by the watchmaker William Freemantle and only decades later named for the naval architect John Scott Russell.

Fig. 7: America's first working steam "locomotive": not only the first practical high-pressure steam-powered vehicle of any sort, but the first amphibious vehicle as well, though this was largely an accident. Evans's workshop was more than a mile inland and fifteen miles from the dredging site on the Schuylkill River, so he added wheels to his paddle-driven dredge, which he had named *Orukter Amphibolos,* a drawing of which is in the upper right corner, and drove it proudly down Philadelphia's Market Street. *Image by permission of the Library of Congress*

lic. But he ran out of money before completing its sequel for steam engineers, and he published it, rather petulantly, under the title *The Abortion of the Young Steam Engineer's Guide.*

Sometimes his petulance verged on a persecution complex: "And it shall come to pass that the memory of those sordid and wicked wretches who opposed [my] improvements, will be execrated, by every good man, as they ought to be now. . . ."

In 1806, however, Evans had opened the factory he called the Mars Works, where he built more than a hundred steam engines and boilers and thousands of components for mill machinery, enough that he ended up dying a rich man in 1819.

Evans was a visionary and a pioneer. But despite his prediction that "the time will come, when people will travel in stages moved by steam engines from one city to another almost as fast as birds can fly," his greatest contribution to the history of steam locomotion was almost incidental: his decision to share the design of his boiler and high-pressure steam engine with his compatriots in Britain. As he wrote in the 1805 *Abortion of the Young Steam Engineer's Guide:*

> In 1794–95, I sent drawings, specifications, and explanations, to England to be shown to the steam engineers there, to induce them to put the principles into practice and take out a joint patent for the improvement, in their names . . . Mr. Joseph Stacey Sampson, of Boston [the same one who received the second patent issued in the U.S.] who carried the papers to England, died there, but the papers may have survived.

The timing is suggestive. In fact, it is a powerful bit of circumstantial evidence for the notion that the papers not only survived, but were read by a Cornish mine engineer named Richard Trevithick.

TREVITHICK'S ANCESTRY, ON BOTH sides, is dotted with members of Cornwall's mining aristocracy. His father and uncle were both captains in some of the region's largest and most profitable copper mines, including the legendary Dolcoath mine, where Richard Sr. built the deep adit in 1765 and constructed a Newcomen engine in 1775. Richard Jr.'s early life, partly in consequence, was slightly atypical, as he was neither the product of a formal apprenticeship, nor was he schooled to be a scientist or engineer; instead, by 1784, he was already working in a relatively senior position at Dolcoath, reporting to his father, and over the next ten years, he practiced his somewhat nomadic craft at half a dozen different mines all across Cornwall, including the Tincroft and Wheal Treasury ("wheal" is a Cornish term simply meaning "mine").

By the 1790s, he was a local hero, partly because he was seen as an adversary of Boulton & Watt. The steam engines provided by the Birmingham firm had improved productivity, but their royalty system generated mostly resentment, since the more copper the mines produced, the more gold their owners owed to Boulton & Watt. So when Trevithick testified on behalf of Hornblower during the lawsuit, he endeared himself to most of Cornwall. Trevithick's appearance didn't hurt. He had grown to be a huge man, at least six feet two, and "Cap'n Dick," as Trevithick was widely known, became the subject of Paul Bunyan-like mythmaking; one story has him throwing a sixteen-pound sledgehammer over the top of a twenty-foot-tall Newcomen engine.

Trevithick was his country's champion not only because of his strength, but his cleverness; in 1797, he invented and built a system that improved upon, and so replaced, the chain of buckets that had been used since Newcomen to pump water out of mines. But the engines that drove them were still built by Boulton & Watt, or under license from them—at least until 1800, when Watt's original patent finally expired. The immediate consequence of the expiration was a dramatic increase in the appeal of the Boulton & Watt designs now that they were available without the Boulton & Watt royalties.

Trevithick, however, wasn't trying to imitate the Boulton & Watt engine, but to replace it. Most especially, like Evans, he wanted to dispense with Watt's separate condenser, but unlike Evans, he didn't know whether it was feasible. This was a basic scientific question, for which Trevithick, an indifferent student (his teachers, though they noted his intelligence, also called him "disobedient, slow, obstinate . . . frequently absent and very inattentive"), had no answer. However, he wasn't shy about using the skills of those with more formal education, and during the Hornblower trial, he had struck up what became a lifelong friendship with Davies Gilbert, the future president of the Royal Society. In 1797, Trevithick asked him to calculate how much power would be lost if instead of capturing the steam in a separate condenser, the engine simply exhausted it into the air.

The answer was Trevithick's real eureka moment. Gilbert explained that with each stroke, the cylinder would lose exactly as much pressure *inside* as the pressure *outside:* 14.7 pounds per square inch at sea level. This would obviously be disastrous for a Boulton & Watt separate-condensing engine, which generated only a little more than ten pounds per square inch inside the cylinder; exhausting the condensation would leave it with no pressure at all.

But what if the pressure inside the cylinder could be increased?

In 1800, five years after Joseph Stacey Sampson had brought Oliver Evans's drawings to Britain, and three years after Davies Gilbert had shown that an engine operating at 60 psi would lose only a quarter of its pressure at each stroke, Richard Trevithick introduced his first high-pressure steam engine at the Wheal Hope copper mine. It used almost precisely the same method Evans used for increasing the steam pressure from the boiler. It is not known whether Trevithick saw the Evans design firsthand or—more plausibly—through William Murdock, his neighbor in the town of Redruth for six months in 1797 (and who, despite his employment by the despised firm of Boulton & Watt, was friendly with Trevithick). The sequence of events, however, is persuasive: Sampson carried Evans's drawings to Britain in 1795, to show to "the steam engineers there." The British steam engineer best known to be working on high-pressure steam for locomotion was William Murdock, who was Trevithick's neighbor when the Cornishman asked Davies Gilbert to give a scientist's opinion on what was, in essence, the practicality of Evans's engine.

The Wheal Hope engine showed that a high-pressure engine was practical; a little more than a year later, Trevithick was ready to demonstrate that it was mobile as well. On Christmas Eve 1801, with apparently no warning, Trevithick appeared on the High Street of the town of Camborne aboard a carriage unlike anything anyone had ever seen before. No drawings survive, though a sketch dated a year later shows a four-wheeled flatbed truck with a vertical steam engine set over the front wheels, a twelve-foot-high boiler over the rear, and the driver (wearing what looks like a top hat) set behind. We do have the testimony of one of the blacksmiths who worked on the castings for the machine, "old Stephen Williams":

> Captain Dick got up steam, out in the high-road, just outside the shop at the Weath. When we get see'd that Captain Dick was agoing to turn on steam, we jumped up as many as could; maybe seven or eight of us. 'Twas a stiffish hill going from the Weight up to Camborne Beacon, but she went off like a little bird. . . .
>
> When she had gone about a quarter of a mile, there was a roughish piece of road covered with loose stones; she didn't go quite so fast. . . . She was going faster than I could walk, and went on up the hill about a quarter or half a mile farther, when they turned her and came back again to the shop. . . .

To the spectators at Camborne Hill wrapped up against the cold that Christmas Eve, the vision of a wheeled vehicle moving uphill without being either pushed or pulled must have seemed something like levitation. For millennia, the force driving wheeled vehicles had always been something external to the vehicle, which meant that the primary traction against the roadway was also generated externally: by a horse's hooves, for example. It was by no means obvious that simply turning a wheel would generate enough traction to pull itself—the first bicycles were still decades in the future—and Trevithick apparently spent the week before his experiment hand-cranking the wheels of a model along cobblestone and dirt roads.

Just as startling to the audience lining the half-mile-long High Street, the engine on board the Camborne carriage made its journey belching smoke like something out of myth. Trevithick's engines, having dispensed with a separate condenser, were thereafter known as "puffers" because their steam exhausted directly to the air. The distinctive clouds familiarly associated with steam engines* had finally been born.

It didn't survive its own infancy; the engine that Stephen Williams rode up Camborne Hill on Christmas Eve was destroyed

* Confusingly so. Steam is actually invisible; the clouds are just evidence that steam has condensed back into water vapor.

before New Year's. Two of Trevithick's drivers celebrating the season in a local pub left the boiler unattended, the water boiled away, and the chamber became hot enough to set fire to the engine, and to the shed in which it was housed. Which didn't, in the end, matter all that much. In March 1802, Trevithick, with help from a fellow Cornishman, the scientist Humphry Davy, applied for and received a patent on the new locomotive.

The market for self-propelled steam engines was still a fraction of that for stationary ones, thousands of which were by then pumping water and operating machinery throughout Britain. One of the more avid users was the Coalbrookdale foundry, then run by the fourth generation of the Darby family, where the boiler for the Camborne Hill locomotive was built; by the end of 1802, the Darbys had hired Trevithick to build a new stationary engine operating at the seemingly impossible pressure of 145 psi.

Their faith in the Cornishman, and in high-pressure steam, was understandable, but it was controversial. The future of steam power was very clearly at stake, and the established powers of British manufacturing had everything riding on the existing Boulton & Watt low-pressure designs. The Birmingham firm's founders had retired in 1800, but their sons were, if anything, more hostile both to high-pressure steam and to Trevithick than were their fathers. No one knew this better than Cap'n Dick; after the boiler of a high-pressure engine exploded in Greenwich in September 1803, Trevithick wrote of his belief that "Mr. B. & Watt is abot to do mee every engurey in their power for they have don their outemost to repoart the exploseion both in the newspapers and private letters." But though the Boulton & Watt public relations campaign could slow the adoption of high-pressure steam engines—and infuriate Trevithick—it could not stop it. In 1803, a Welsh ironmaster named Samuel Homfray, jealous of his competitors at Coalbrookdale, invited Trevithick to bring his magic across the Bristol Channel from Cornwall to Wales.

Industrialization in Wales, as in England, had been partly a function of geology, and the great ironworks that had grown up around the town of Merthyr Tydfil in the 1760s were testimony to the generous local supplies of hematite, limestone, coal, and water, the key ingredients for a furnace-and-forge economy. Just as rich was the supply of ironmongers—four large ironworks, each competing with one another to meet the ever increasing demands of British industry. Competing, and also cooperating: the four ironworks each used the same route to transport their goods to Cardiff and the coast: the twenty-four-mile-long Glamorganshire Canal, which they had built, owned, and, in theory, shared.

They did not, however, share it happily. The owner of the Cyfarthfa ironworks, Richard Crawshay, was also the majority shareholder in the canal, and as a result, barges hauling Cyfarthfa iron were granted preferential treatment. His partners, the owners of the Dowlais, Plymouth, and Penydarren works, were angry enough to build a parallel route for their iron, a horse-drawn railway. One of them was Samuel Homfray of Penydarren, Richard Trevithick's new employer.

Homfray's original objective in hiring the Cornishman seems a bit unformed in retrospect. It is certain that he wanted a new

Fig. 8: The Penydarren engine, a replica of which is still on display at the National Maritime Museum at Swansea, was as important in its way as either Newcomen's 1712 Dudley Castle pump, or Watt's 1776 New Willey engine, or even the Stephensons' *Rocket*. *National Railway Museum / Science & Society Picture Library*

steam engine to run the hammer in Penydarren's forge, but unlikely that he had already thought about replacing the horses pulling his iron-filled carts. For one thing, the existing railway was built for horses rather than locomotives; the rails were set into concrete stones set four feet apart, and had no crossties in between to trip up the horses. Also, the grade was extremely gentle; on the way from Merthyr to the wharf at Abercynon, the railway dropped only one foot for every forty-five traveled, putting less strain on the horses both going and returning.

In the event, something provoked Crawshay to bet Homfray five hundred guineas that no steam locomotive could do the job of the horses. To win, Trevithick needed to build an engine capable of hauling ten tons of iron ore from Merthyr to the wharf, nine and a half miles away.

The Penydarren locomotive is practically an encyclopedia of innovation. As protection against explosion, it used one of Trevithick's cleverest inventions, the so-called "fusible plug," a small lead cylinder inserted into a predrilled hole in the wall of the engine's boiler—a hole that, in a properly operating engine, would always be underwater. If, however, the water level in the boiler were to fall low enough to become dangerous, the heat would melt the lead plug, allowing the steam to blow out the fire. The Penydarren engine also incorporated a U-shaped fire tube, a return flue that carried the air heated by the furnace from one end of the boiler and back again, which put at least twice as much surface area in contact with the water. Even more important, it didn't just exhaust the spent steam into the air, but used a chimney that, in Trevithick's own words, "makes the draft much stronger"—that is, the exhaust steam, hotter than the surrounding atmosphere, rose. By doing so, it pulled more oxygen into the furnace, raising its temperature and increasing the efficiency of the heat engine itself.

In other respects, it was a bit of what a later generation of engineers would call a kludge. Trevithick was obliged to build an engine that would serve Homfray's forge as a steam hammer, whether or not it worked to transport his ore, and it was therefore cobbled together from pieces intended for different functions. Its piston operated like a slide whistle, driving only the two wheels on the engine's left side and conserving momentum with an enormous flywheel set behind them. And, on February 21, 1804, it worked—sort of:

> . . . yesterday we proceeded on our journey with the engine, and we carried ten tons of iron in five wagons, and seventy men riding on them the whole of the journey . . . the engine, while working, went nearly five miles an hour; there was no water put into the boiler from the time we started until our journey's end . . . the coal consumed was two hundredweight.

And, sort of, it didn't. The problem lay less with the locomotive than with the rails, which cracked like twigs. The engine, whose five and a half tons were distributed over only four wheels, and with only two of them driving, put an unanticipated lateral strain on the railway. Though Trevithick would try again, on a coal railway in Newcastle in 1805, the rail problem remained unsolved; even in 1808, when Trevithick demonstrated his "Catch-Me-Who-Can" locomotive on a half-mile oval near Gower Street in London for a shilling a head, it was regarded more as a circus act than as any useful industrial advance.

Trevithick's engine, the first driven by high-pressure steam, earned him a considerable claim on the title "father of railways," but the birth of steam locomotion was still a decade or so in the future. More important, though less romantic, was another of Trevithick's innovations, one that was nearly as large an improvement over the first high-pressure design as that had been over the Boulton & Watt separate condensing engine—indeed, as big an improvement as Watt's separate condenser was over Newcomen's original atmospheric engine.

For nearly a decade, Trevithick's high-pressure engines had been making significant inroads into the dominant position of Boulton & Watt in Cornwall's mines. By 1812, he determined to displace them once and for all. In a pump built for the Wheal

Prosper mine in Cornwall, Trevithick modified his existing high-pressure steam design so that instead of exhausting the condensed steam directly to the atmosphere, as with the Penydarren and Camborne Hill engines, he allowed it to expand into a lower-pressure chamber first. In the new engine, the pressure on the piston came from *both* the expansive property of high-pressure steam on top of the piston *and* the atmospheric pressure on the chamber once the steam has been condensed. Steam flowed into the top half of the cylinder and pushed the piston down some distance, at which point a valve closed and the steam expanded to fill the now smaller volume. Trevithick, in comparing an early model to a Boulton & Watt atmospheric engine, discovered that he could produce 40 psi using one-third the coal that the atmospheric engine needed to produce 4 psi. Even more innovative, the new engine's boiler lay horizontally, which allowed the fire tube to run through its middle, heating the water both efficiently and to high pressure. "My predecessors," Trevithick said, "put their boilers in the fire; I have put the fire in the boiler." The result, in 1812, was the first really successful "Cornish engine."

It was certainly successful as measured by the still-in-use benchmark of "duty," which measured the pounds of water raised one foot by a bushel of coal. A high-performing Newcomen-style engine typically performed in the neighborhood of 5,000 pounds; Smeaton's many improvements nearly doubled that number—to 9,600 pounds—without changing the basic design, and a 1778 Watt engine, with separate condenser, achieved a duty of 18,900 pounds. By 1812, Trevithick was boasting of 40,000 pounds, which is likely an exaggeration, but an objective report of three Cornish engines at the Dolcoath mine reported 21,400, 26,800, and 32,000 pounds in 1814. By 1835, another Cornish engine achieved a duty of 100,000 pounds.

However, efficiency, as measured in duty, was not everything. The price of the Cornish engine's dramatic achievement was that its multiple chambers and valves demanded an unforgiving level of both precision and maintenance. Without either, they were more subject to breakdowns—and to the purchaser of a steam engine trying to make delivery of a scheduled amount of cotton, produce a quantity of iron, or pump water, it mattered little to have the most efficient steam engine if it was out of commission for two days a week. As a result, it is the last advance in steam power with Trevithick's name attached to it.

Instead, like a homing pigeon, he returned to his origins: precious metals mining. Trevithick became obsessed with reopening the silver mines of Cerro de Pasco in Peru, which had once been among the richest of Spain's possessions in the New World. Trevithick convinced himself that he would be able to make the Peruvian mines profitable once again, and he left Britain planning to do so in October 1816, arriving in February of the following year.

His timing could have been better. By 1817, most of South America was in rebellion against Spain; the month before Trevithick arrived in Peru, the Argentine general José de San Martín had crossed the Andes into Chile and was preparing to head north. A month before that, Simón Bolívar had returned to Venezuela from Haiti. Though Peru would remain under Spanish control for another five years, Trevithick's engines (he had shipped four pumping engines and four winding engines ahead of his arrival) were still in their crates when his romantic soul got the better of him and he joined the rebellion. While in Caxatambo, Peru, he even designed a new carbine for Bolívar's army, but when the city was occupied by the Spaniards in 1818, Trevithick was forced to flee north to Costa Rica, leaving an estimated £5,000 in ore and uncounted more pounds' worth of lost equipment.

Trevithick's South American adventure carries an almost unwieldy tonnage of symbolism: a representative of the dominant world power of the nineteenth century caught in the collapse of the dominant one of the sixteenth. Even more pointedly, it offers a high-contrast picture comparing history's two longest-lasting approaches to the very idea of wealth: wealth as technology versus wealth as precious metals. Whatever meaning is retrospectively poured into it, however, the experience as Trevithick lived it seems to be less about the metaphorical war between two notions of political economy than about the real thing. Evading Spanish patrols in the Nicaraguan jungles, which the Cornish inventor was forced to traverse on foot, destroyed whatever romance the rebellion still offered. Trevithick's journey, which included a dozen hair's-breadth escapes, the deliberate capsizing of his boat by an offended traveling companion, and bouts of illness too frequent to count, found him arriving at last in Cartagena, Colombia, exhausted, sick, and broke—in his own words, "half-drowned, half-dead, and the rest devoured by alligators."

There his story took an unlikely turn. In Cartagena, he met the son of an old friend, who lent him £50 for his fare home. When Trevithick finally returned in 1827, he had nothing on his person to show for a decade in South America but two compasses—one for drafting, the other for navigating—a pair of silver spurs, and his gold watch.

Aficionados of dramatic coincidences could, however, take some comfort in the name of the man who paid for Richard Trevithick's ticket home. He was Robert Stephenson, of Newcastle, and along with his father, George, is Trevithick's only serious competitor for the title of "father of railways."

THE DEPICTION OF GEORGE STEPHENSON by Samuel Smiles, the prolific biographer and self-help author* who did more than anyone else to establish the heroic archetype for British inventors, is a textbook example of self-discipline and deferred gratification. His first job was as a picker: a laborer whose entire job was separating coal from the stones that accompanied it from mineshaft to colliery. Soon enough he was working as an assistant fireman, then as the "plugman" operating a set of valves on the steam-driven pump at another collier's. When he turned twenty, he was appointed as the brakeman, responsible for main-

* Literally; in addition to biographies of Watt, Smeaton, Maudslay, Dudley, Boulton, and dozens of other inventors and engineers, he also wrote, in 1859, the worldwide bestseller titled *Self-Help: With Illustrations of Character and Conduct*.

taining the winding mechanism that pulled the coal-filled buckets from the work area of the mine.

None of this permitted much time for education, even of the practical sort, and in a previous century, Stephenson might have simply become a superbly reliable artisan. By the end of the eighteenth century, however, Britain—particularly outside London—had spent decades making heroes out of onetime laborers who had become wealthy by acquiring and producing useful knowledge. So inspired, Stephenson taught himself to read and write and hired someone to teach him the rudiments of arithmetic. By 1801, the ambitious twenty-two-year-old brakeman took on additional work moonlighting as a watch repairman; two years later, now a father, he took charge of the Boulton & Watt steam engine driving the wheels of a Scottish spinning factory that had grown prosperous making cloth for the military uniforms needed for the war against Napoleon. This did not exempt Stephenson from service himself, nor did the birth of his son, Robert, in 1803; the following year he was drafted for the militia, and went into debt paying for a substitute to serve in his place. The same year, he returned to the Killingworth pit mine, where he taught himself mechanical engineering by spending his one day off each week dismantling and reassembling the colliery's steam engine.

In 1811, Stephenson, like Watt thirty years before, made his first real mark on the world repairing a Newcomen engine, this one an old model at one of the Killingworth pits. For the job he was paid £10, and far better, was hired to manage all the engines owned and operated by the collieries of the so-called "grand allies," a group of aristocratic investors that owned the Killingworth pit, at the impressive salary of £100 per year—the equivalent of more than $100,000 in current dollars. Stephenson's salary was an insurance policy for the stationary engines on which the colliers depended; it rapidly turned into an investment in an entirely new industry. In 1814, Stephenson built his first locomotive to transport coal at Killingworth: the *Blucher,* a giant step toward practical steam locomotion.

. . .

THE TWO GREAT PROBLEMS in harnessing steam power for transportation were, broadly speaking, both a function of weight. The first one—increasing the engine's power-to-weight ratio—was addressed, if not solved, by the realization, by both Evans and Trevithick, that more heat meant more pressure and therefore more work. That notion, which is obvious in retrospect but was revolutionary at the time, practically demanded a whole series of "micro-inventions" intended to turn up the dial on steam boilers: return flues, for example, which not only "put the fire in the boiler" but increased exponentially the area heating the water. Even more important, exhausting the steam through a chimney located above the furnace created a draft—a "steamblast"—that raised the heat even further.

The other weight problem was the one that licked Trevithick at Penydarren: The tracks on which the locomotive ran were just not able to survive the tonnage traveling over them. Driving a five-ton steam locomotive over rails designed for horse-drawn carts was only slightly more sensible than driving a school bus over a bridge made of wet ice cubes. In both cases, it's a close call whether the vehicle will skid before or after the surface collapses.

This is why all of the dozens of inventors attempting to put steam on the move were obsessed with the durability and traction of the surface holding their vehicles; for centuries, the rails originally designed for horse transport had been made of wood, occasionally reinforced with iron edging. Not until 1767 did the Darbys of Coalbrookdale begin casting iron rails for wagonways, which made them far stronger; within twenty years some unknown innovator had added an arched rim, or lip, to prevent wheels from slipping off.

The rims, or flanges, were fine for keeping the wheels from moving laterally, but they did nothing to increase traction—a real challenge for smooth iron wheels on smooth iron rails. In 1811, John Blenkinsop, an employee of another collier located in the city of Leeds, patented a Trevithick-style engine with a cogged driving wheel, and accompanied this with a new sort of track, this onc made of cast iron with an edge rail carrying a toothed rack. The cog-and-rack not only eliminated any possibility of skidding, it transmitted five times the force of Trevithick's original engine, and Blenkinsop-style engines remained popular through the 1820s, despite the enormous cost of producing miles of what were essentially horizontal iron gears.

Two years later, a civil engineer named William Chapman applied for and received a patent for an engine that propelled itself by hauling itself along a chain; even odder, William Brunton also managed to patent his "Brunton Mechanical Traveler" that "walked" the locomotive along by operating mechanical feet driven via a complicated series of levers and linkages. Slightly less eyebrow-raising, in 1815, William Hedley's "Wylam Dylly" engine tried to solve the excess weight problem by doubling the axles from two to four, thus distributing the weight over a larger area.

Stephenson's locomotive, which made its maiden journey on July 25, 1814, took a different approach. The engine driving the *Blucher* (named for the Prussian general who would pull the Duke of Wellington's chestnuts out of the fire at Waterloo almost exactly a year hence) incorporated an early version of the blastpipe: a vertical tube with a narrowed exit that carried the exhaust into the chimney, creating a draft, just as Trevithick had more or less accidentally discovered a decade before. It also ran on a reversed version of the most popular track design, putting a flanged wheel on a smooth rail. Two years later, Stephenson, in collaboration with the ironmonger William Losh of Newcastle, produced, and in September 1816 jointly patented, a series of improvements in wheels, suspension, and—most important—the method by which the rails and "chairs" connected one piece of track to another. Stephenson's rails seem mundane next to better-known "eureka" moments, but as much as any other innovation of the day they underline the importance of such micro-inventions in the making of a revolution. For it was the rails that finally made the entire network of devices—engine, linkage, wheel, and track—work.

Stephenson's working life* marks the point in the development of steam technology when the value of what economists call "network effects" finally overtook the importance of any individual invention, however brilliant. Setting the distance between the smooth tracks on which the *Blucher* traveled at four feet eight and a half inches was arbitrary—that was the width of the Killingworth Colliery wagonway—but its specific width was irrelevant. The value of any standard is not its intrinsic superiority, but the number of people using it. Like the famous example of the QWERTY keyboard, the Stephenson gauge became the world standard, and it is still the width used on more than 60 percent of the world's railroads.

Of course, simply laying rails a particular distance apart does not make for a monopoly unless others follow. And others weren't about to follow Stephenson's lead until they were persuaded that there was some advantage to it, in the form of either increased revenue or lower costs. To a proprietary line, such as the ones that connected coal mines with ports, the advantage of a standard wasn't all that obvious; as at Killingworth, it was frequently more economical to use existing wagonways and roads than to redesign them to a new standard. The same didn't apply to so-called "common carriers," who needed, by definition, to accommodate rolling stock they didn't own, or to travel on railways they didn't build. The first common carrier to realize this, the Stockton and Darlington Railway, was, not at all coincidentally, one of George Stephenson's employers. But by far the most important one was the one intended to connect the cities of Manchester and Liverpool.

It is almost indecently tempting to place the Liverpool & Manchester Railway at the climax of the entire history of British in-

* Not all of Stephenson's historically significant inventions were associated with railroads, or even steam. His invention of a safety lamp, one that placed a barrier such as metal gauze between the candle and surrounding gas practically saved the deep coal mining industry. Stephenson's eponymously named "geordie" was virtually simultaneous with a similar one invented by the Cornish chemist, and onetime partner of Richard Trevithick, Humphry Davy; the dispute over primacy continues to this day.

dustrialization. The first temptation is posed by Manchester itself, which was, when George Stephenson took on the job as chief engineer of the proposed railway in 1825, the most "industrial" spot in all England, with all that implied: "a town of red brick," in the words of Charles Dickens, "or of brick that would have been red if the smoke and ashes had allowed it."

The reason, of course, that those bricks were covered by smoke and ash was that the city was home to the world's largest textile manufacturers, factories that used coal to turn cotton into clothing. Richard Arkwright's mills, which gave the city its nineteenth-century nickname of "Cottonopolis," had become so successful that the choke point for the industry's growth was no longer technological imbalance (the difference between efficient spinners and inefficient weavers, for example) but transportation. Manchester was making cotton faster than it could ship it, and Britain's canal system, even with its sophisticated locks, was less and less able to handle the load, which was easily exceeding a thousand tons of cargo daily: raw cotton in, finished goods out. So much cloth was being made in Manchester, in fact, that by 1800, the port of Liverpool on the River Mersey was the world's most important; less than eighty years old, it handled more than a third of all the world's trade. The need for a railway to connect Manchester's mills with the port city had become urgent.

It is metaphorically satisfying to talk about threads being woven together when talking about cotton, but the thread that mattered to the Liverpool & Manchester Railway was made of iron: thirty miles of it, smelted, forged, and wrought in ironworks like Coalbrookdale on the Severn, and laid down as rails between the two cities that were now producing, in their mundane way, more wealth in a year than the entire Roman Empire could in a century.

But while there were clearly massive financial incentives for building *some* kind of railway between the factory and the port, the railway's directors were uncertain that a *locomotive* railway was the best option. Some of the investors and directors in the enterprise were promoting the use of rope cables to haul boxcars full of cotton the entire thirty miles, using stationary engines roughly every mile and a half. Others wanted different kinds of locomotives (though no one, happily, was arguing on behalf of Brunton's mechanical "walker"). After much to-ing and fro-ing, it was decided to settle the problem with a contest.

On May 1, 1829, the Liverpool & Manchester Railway ran an advertisement in the *Liverpool Mercury* inviting "engineers and iron founders" to submit plans for locomotives to compete for the winning design. The offer of a £500 prize, the equivalent in average earnings of more than $500,000 in 2010, brought the crackpots out in force. The treasurer of the Liverpool & Manchester, Henry Booth, described the applications:

> From professors of philosophy down to the humblest mechanic . . . [from] England, America, and Continental Europe. Every element and almost every substance were brought into requisition and made subservient to the great work. The friction of the carriages was to be reduced so low that a silk thread would draw them, and the power to be applied was to be so vast as to rend a cable asunder. . . . Every scheme which the restless ingenuity or prolific imagination of man could devise was liberally offered to the Company. . . .

The oversupply of perfect vacuums and perpetual motion machines was in part a testimony to the utter transformation of British cultural attitudes toward innovation over the preceding century. By the 1820s, the Patent Office was approving nearly three hundred new inventions annually, and rejecting thousands. In the event, the Liverpool & Manchester had made the conditions for entry fairly strict: entries had to be mounted on springs, weighing no more than six tons including water (if on six wheels) or four and a half tons (if on four); they must operate at between 45 and 60 psi, while being prepared for a test at up to 150 psi; they must consume their own smoke (to keep the route as clear of ash as possible; this effectively required the engines to burn coke rather than coal); and they were required to pull a gross load of twenty tons at ten miles an hour back and forth along a mile-and-a-half course forty times, reproducing the sixty-mile round trip between Manchester and Liverpool.

The stipulations eventually weeded out all but five applicants, only three of which could be called serious. One of the others never actually made it to the starting line, and the other—the *Cycloped,* whose source of propulsion was a horse trotting on a treadmill and which was only allowed to compete because its designer was on the railway's Board of Directors—proved good for nothing more than comic relief.

Two of the remaining three competitors were joint favorites to win the prize: the *Sans Pareil,* built by Timothy Hackworth, master mechanic of the Stockton and Darlington Railway (and therefore George Stephenson's former employer), and the *Novelty,* the creation of a former Swedish army officer now living in London, John Ericsson.

The third, entered by Henry Booth and George Stephenson and to be built by his son Robert—Richard Trevithick's rescuer, and an even more skilled engineer than his father—was *Rocket.*

Between May and September of 1829, Robert Stephenson—who had promised his father, "Rely upon it, locomotives shall not be cowardly given up. I will fight for them until the last. They are worthy of a conflict"—labored at his workshop in Newcastle-on-Tyne to construct the world-changing locomotive. While it incorporated a key design feature suggested by Booth (the multitube boiler, about which more below), every other innovation contained in the final entry was the work of Robert, who had explicitly identified four areas for potential improvement in the final design: transmission of the largest amount of power from the pistons to the wheels; preservation of the greatest amount of traction between wheels and track; minimizing the loss of heat between boiler and cylinder; and maximizing the amount of heat within the boiler itself.

Those innovations are the reason that any list of the most significant engines in locomotive history always includes the Stephensons' entry at Rainhill, the site of the competition's final trials. First were its mechanics: the way it transmitted the reciprocating motion of its pistons to its wheels. The "premium engine," as the two Stephensons referred to it, used two pistons set at a 45-degree angle above the front axle, each one attached to a slip eccentric, which is a sort of linkage in which a disk is attached to an axle but offset "eccentrically" (essentially a simpler, and more efficient, version of the sun-and-planet gear). One set of slip eccentrics turned the reciprocating motion of the pistons into rotation, while another set worked in reverse, opening and closing the steam valves as the engine cycled.

To increase the amount of traction between wheels and track, Stephenson and his assistant William Hutchinson calculated the optimal arrangement of weight over the wheels and determined to use the engine to operate only the front wheels; it was far more efficient, both in tractive power and durability, to drive only two large (4′8″ diameter) wheels and use the back wheels (with a diameter of 2′6″) for balance.

Fig. 9: Little though it resembled the great locomotives of the nineteenth century, *Rocket* pioneered virtually all of their engineering innovations, from the high "blastpipe" chimney to the multitube firebox to the slip eccentric gears on the driving wheels. *National Railway Museum / Science & Society Picture Library*

But the truly revolutionary significance of the engine was its boiler design. Twenty years before Rainhill, Oliver Evans had demonstrated that raising the boiler's heat by doubling the amount of fuel increased the engine's power by at least ten times; Richard Trevithick had goosed up the heat in his boiler with a U-shaped return flue. The principle was, in retrospect, obvious: Since the water was heated by conduction with the chamber containing heated gas, increasing the surface area of the chamber would transmit more of that heat to the water surrounding it. Robert Stephenson was just about ready to take that principle to its logical conclusion.

Rocket's boiler did not have a single flue, even a U-shaped one. Instead, as suggested by Henry Booth, twenty-five copper tubes, each three inches in diameter, were fitted into a firebox inside a water jacket, with somewhat wider copper tubes connecting them to the barrel of the six-foot-long, three-and-a-half-foot-diameter boiler. The cylinders exhausted their steam into two blast pipes inside the chimney, whose slightly narrowed openings guaranteed a powerful draft of air. Robert Stephenson spent the entire month of September testing to ensure that the boiler and cylinder were reliably steamtight to the point that they could handle up to 150 pounds of pressure per square inch. It finally passed Stephenson's inspection only the day before it left his Newcastle workshop and was placed on a series of horse-drawn carts for the 120-mile journey to the Rainhill course, ten miles east of Liverpool.

The first day of the trials, October 6, was largely a day for demonstration, as each competitor tried the course without hauling the weight required by the contest's rules. *Novelty,* at two and a half tons, the lightest of the three remaining entrants, was by far the fastest. Using two vertical cylinders to drive a crank attached to the leading axle, it was also, by general consent, the prettiest engine in the competition (painted royal blue, with its boiler and water tank covered in polished copper), and it was made the early favorite, a position it improved on the following day, when, hauling more than eleven tons, *Novelty* easily hit a speed of 20 mph.

On October 8, the final specifications for the contest were published: Each engine was no longer required to haul twenty tons, but a load of three times its weight, including the water in its boiler, with allowance made for the engines—*Novelty* and *Sans Pareil*—that hauled water in the locomotive rather than in a separate tender. Though the entrants were to have competed in the order of their "race cards"—*Novelty,* then *Sans Pareil,* then *Rocket*—the first two needed last-minute repairs, and *Rocket* went first.

Rather surprisingly, given the historical significance and number of spectators, no one knows who actually drove *Rocket* on its October 8 debut at Rainhill. Robert McCree, from the Killingworth Colliery, had driven it during testing, but at least one report suggests that he was, like Robert Stephenson, only a passenger (and possibly a fireman, loading fuel into the firebox). If so, the driver could only have been George Stephenson himself, and he, like *Rocket,* covered himself in glory, along with coal dust. It took only fifty minutes for the fuel (like the other competitors, *Rocket* used cleaner-burning coke, to "consume its own smoke") to bring the pressure in the boiler up to the required 50 psi from a cold start, and by 10:00 A.M., the engine was on its way.

And so it continued. Aware that the rules mandated an average speed of 10 mph, the Stephensons kept their pressure well below its maximum for the first back-and-forth laps. It took a bit more than six minutes, at an average speed of around 15 mph, to complete the first mile and a half: just about the pace of a good twenty-first century fifteen-hundred-meter runner. By the tenth lap, the engine was moving closer to 20 mph, but not until the last lap did the Stephensons open up the steam regulator and let *Rocket* fly. When they passed the grandstand at the eastern end of the course, Rocket was pulling its twenty tons at more than 30 mph, all while consuming "only" a little more than 200 pounds of fuel an hour. Thousands of spectators rushed the finish line to cheer the Stephensons on their triumph.

The rest of the competition was something of an anticlimax. *Novelty* didn't get a chance to compete until Saturday, October 10. The same high power-to-weight ratio that had made it such a fan favorite four days earlier allowed it to race off at what must have seemed magical speed, completing its first mile in less than two minutes. Before its second, however, a blowback from the engine's furnace burst the bellows used to create chimney draft. The explosion ended *Novelty*'s day. On its next run, the favorite managed only one lap before another pipe exploded; since this was the pipe that fed the boiler, the resulting detonation ended with "the water flying in all directions." When the boiler gave out, Ericsson gave up.

Sans Pareil did perform brilliantly. The heaviest of the three finalists, it pulled a full twenty-four tons at better than 15 mph. But not for long. After twenty-two and a half miles, its boiler, rather embarrassingly, ran dry, melting the fusible plug that stopped it cold. The reason was its enormous consumption of fuel: nearly 700 pounds per hour.* The victor, by acclamation, was the Stephensons' *Rocket.*

IT'S NOT NECESSARILY OBVIOUS that the Rainhill Trials mark the moment in history when the steam revolution became finally, and utterly, inevitable. One year later, the Liverpool & Manchester Railway opened for business, with eight Stephenson-built locomotives traveling on Stephenson's standard-gauge track before luminaries that included the then prime minister, the Duke of Wellington,† but the conflicts over the proper use of steam power didn't vanish. To the end of his life, Stephenson fought a running battle with an even more famous engineer, Isambard Kingdom Brunel, over the latter's preference for an "atmospheric railway system" operated by stationary engines. Brunel, the son of Marc Brunel, of the Portsmouth Block Mills, even designed the Great

* Hackworth never did accept his loss at Rainhill, and he and his supporters argued that the boiler failure was actually sabotage; perhaps imprudently, Hackworth had ordered it from Robert Stephenson's workshop in Newcastle-on-Tyne. In the event, his accusation was dismissed, and the Liverpool & Manchester Railway ended up buying *Sans Pareil.*

† The inaugural day for the Liverpool & Manchester is famous for the death of Liverpool MP William Huskisson, who was run down by *Rocket.* Just as widely reported, and far more lauded, was the heroic dash George Stephenson made in *Rocket* to the nearest hospital, during which he averaged 36 mph for fifteen miles.

Western Railway to run on a gauge nearly three feet wider than Stephenson's (though he soon discovered the impossibility of overcoming an early monopoly advantage).

There is, after all, something as arbitrary about ending the story of the steam revolution at Rainhill in 1829 as beginning it in first-century Alexandria. Unfortunately for historians, if not for history, such convenient end points are as capricious as the textbook dates for the Industrial Revolution itself, which the careful reader will remember were originally matched as a lecture hall convenience to the regnal years of George III. One might just as well have decided that the story ended in 1819, the year that James Watt and Oliver Evans died—and, coincidentally, the year of the first steamship crossing of the Atlantic, by the American-built *Savannah.* Or 1824, when Sadi Carnot finally explained the thermodynamics of steam power. Or 1838, when I. K. Brunel's *Great Eastern* connected a steam railroad with a true transatlantic steamer (the *Savannah* was really a three-masted sailer, with paddlewheels added).

The reason for ending with Stephenson's triumph nonetheless seems persuasive. Rainhill was a victory not merely for George and Robert Stephenson, but for Thomas Savery and Thomas Newcomen, for James Watt and Matthew Boulton, for Oliver Evans and Richard Trevithick. It was a triumph for the ironmongers of the Severn Valley, the weavers of Lancashire, the colliers of Newcastle, and the miners of Cornwall. It was even a triumph for John Locke and Edward Coke, whose ideas ignited the *Rocket* just as much as its firebox did.

When the American transcendentalist Ralph Waldo Emerson met Stephenson in 1847, he remarked, "he had the lives of many men in him."

Perhaps that's what he meant.

EPILOGUE

THE FUEL OF INTEREST

LEADVILLE, COLORADO, AT AN elevation of 10,152 feet, is the highest city in the United States, though the term "city" is generous; fewer than three thousand people live there, most of them directly or indirectly supported by tourism. Leadville, like many places in the American West, trades on its history, and it has more to trade on than most. Leadville was where Doc Holliday escaped after the legendary gunfight at Tombstone's O.K. Corral, and the hometown to which the "unsinkable" Molly Brown returned after surviving the sinking of the *Titanic*. But most of the local color comes from local mines, from which millions of dollars in gold, and especially silver, were extracted in the last decades of the nineteenth century—enough, in fact, that Leadville is home to the National Mining Hall of Fame and Museum.

It is also where, on October 11, 1962, the last regularly scheduled steam locomotive in the United States departed on its fourteen-mile trip to Climax, Colorado. The engine—#641 on the books of the Colorado & Southern Railway—used a multitube boiler fed by a blastpipe, just like *Rocket.* Just like *Rocket,* it ran on standard-gauge track, four feet eight and a half inches wide, just big enough to carry coal down the old wagonway at Killingworth Colliery.

And just like *Rocket,* engine #641, built in 1906 by Philadelphia's Baldwin Locomotive Works, is practically an encyclopedia of engineering innovations, hundreds of them invented after the Rainhill Trials. In the 1840s, locomotives worldwide adopted a different linkage arrangement—the so-called "valve gear," also a George Stephenson patent—but still connected pistons to wheels using slip eccentrics. By the time engine #641 was being designed, even Stephenson's valve gear was supplanted by a different and superior version invented by the Belgian engineer Egide Walschaerts. *Rocket*'s angled pistons were replaced by horizontal ones. Fireboxes moved forward and back, wheel arrangements changed. Superior pressure gauges replaced the nine-foot-tall mercury tube used at the Rainhill Trials; air brakes were introduced, and then were replaced by ones using vacuum—necessary for stopping twentieth-century locomotives that could be one hundred and twenty feet long and weigh five hundred tons even without freight. By 1900, railroad track in Great Britain covered more than 48,000 miles; in Europe, more than 65,000. The United States, with its enormously greater territory, had laid 193,000 miles of track on its way to a 1930 peak of 230,000.

As on land, so at sea. From the end of the 1860s through the 1920s, oceangoing ships turned their screws using engines with three or more cycles of expansion that were just a logical extension of the original compound engine, each cylinder using exhausted steam at lower temperature (and therefore pressure) in a greater volume of space, usually by increasing cylinder diameter. Since steam engines need fresh water, using the final stage of condensation for the boilers made possible the great steamships of the early twentieth century. Inventions created to move freight uncovered a new and highly profitable business in transporting large masses of people.

Innovation in stationary steam engines was, if anything, even more dramatic. On March 10, 1849, the American George Henry Corliss received a U.S. patent for "certain new and useful improvements in Steam-Engines," and he was being modest. The Corliss engine incorporated a rotary valve (and a version of Watt's centrifugal governor) to offer variable control to the steam and exhaust ports in the cylinders, which resulted in a massive increase in efficiency and some extraordinarily massive engines: The Corliss Centennial Engine, forty-five feet tall, with a flywheel diameter of thirty feet, produced more than 1,400 horsepower and operated virtually every moving part at the Philadelphia Centennial Exposition of 1876.

It wasn't until twenty years after Rainhill that science finally caught up with steam engineering. In 1851, a Scottish physicist (and onetime railway engineer on the Edinburgh & Dalkeith Railway, where his father was superintendent) named William John McQuorn Rankine published a paper demonstrating that the theoretical efficiency of steam engines—of any heat engines—could be precisely measured by establishing the upper and lower working temperatures of the system. This tempted John Ericsson (*Novelty*'s designer, who should have known better) to build what would have been a perpetual motion machine: an attempt to use the heat lost in the exhaust over and over again in a huge machine with four cylinders, each fourteen feet in diameter. Ericsson was not the first, but very nearly the last, inventor to fail to understand that heat, once converted into work, is no longer available as heat. Though Ericsson's engine could save fuel (it was also known, appropriately enough, as an *economizer*), it could not use the same energy twice.*

More practically, two New Yorkers, John Allen and Charles Porter, revolutionized the ability of steam engines to deliver rotary power, adapting Watt's governor to spin at very high speeds, which an 1858 article in *Scientific American* declared "if not absolutely perfect in its action, is nearly so, as to leave in our opinion

* In 1861, Ericsson, who had emigrated to the United States in 1839, designed and built the revolutionary ironclad steamship the *Monitor*.

nothing further to be desired." This turned out to be critical, since by the 1880s the ability of reciprocating steam engines to drive a rotary gear at high speed had acquired a new purpose: the production of electricity.

The first electric generators—coils of wire, usually copper, spinning between the poles of a magnet—required startlingly high speeds of rotation; an 1887 machine ran at more than 1,600 revolutions per minute. All that spinning produced a lot of power, and demanded a lot. The biggest reciprocating steam engines ever built were ordered in 1899 for the electrical systems of New York City's subway system. The ten-thousand-horsepower monster weighed in at more than seven hundred tons, with a series of thirty-foot-high cylinders driving an alternator that, all by itself, weighed 445½ tons. Even these massive engines were soon replaced by steam turbines, whose thermal efficiency is at least twice that of the best reciprocating engine—turbines convert up to 80 percent of heat energy into work, as opposed to less than 30 percent in a Cornish engine.

Steam turbines produce more than three-quarters of the world's electricity, but they don't drive the successors to *Rocket* and engine #641. Diesel-electric trains, like automobiles and propeller-driven aircraft, use internal combustion—steam engines, because their furnaces boil water in a chamber outside the cylinder, are external combustion machines—for the same reason that high pressure was needed to put steam on the move in the first place: a superior power-to-weight ratio. The conversion from steam to diesel-electric railroading, in which diesel engines drive electric traction motors, began in the early twentieth century and was completed in most of North America, Europe, and the United Kingdom by the 1960s. The significance of this is less than meets the eye. Steam locomotives may be harder to find outside museums (though not impossible; they remain popular in Asia and Africa, and even on some narrow-gauge lines in western Europe), but steam power is very much a going concern. Though it is now produced mostly by turbines instead of pistons, and delivered not by connecting rods but by copper wires, the world still burns a lot of coal to turn water into steam.

There is no doubt that the thermodynamic gradient between liquid water and steam changed the world, or that its discovery marks one of the most important turning points in history. It is not, however, the "most powerful idea" of this book's title. To find the really big turning point in history that we associate with steam, and industrialization, we have to look elsewhere.

The classical Greeks, in their dramas, called the turning point the κρισις (which has retained its original pronunciation and definition—"crisis"—for three millennia). The word, derived from a root meaning "I decide," refers to a moment after which the protagonist's fate is changed forever. It seems pretty clear that we know when one of those before-and-after moments occurred. And we know where: three centuries back, in Britain and Britain's colonies.

To many, the before-and-after snapshots do not make a happy comparison; contemporary opinion has become decidedly mixed about humanity's leap into the age of fossil-fueled machinery.

And not just contemporary opinion; the early romantic—in an English literature class, it would be capitalized—appeal of the new science didn't survive the first decades of the nineteenth century. Wordsworth, Thackeray, Carlyle, and later Charles Dickens, John Ruskin, and William Morris were uniformly appalled* by the impact of machines on (take your pick) the rural countryside, the traditional family, the joy of craftsmanship, or any combination thereof.

Their prescience did not, however, extend to the impact of carbon on the planet's atmosphere.

It's a cheap shot to call the movement to reverse human-caused global warming a descendant of Carlyle's sneers about what he called a "Mechanical Age." And it wouldn't much matter if it were. If in the beginning the known costs of industrialization had included irreversible climate change in the form of melting

* Reams of Ph.D. theses have explored why industrialization was greeted with such a different attitude on the other side of the Atlantic, where Emerson, Whitman, and Hart Crane could all celebrate the new nation's bridges and railroads. It's not unrelated to America's early displacement of Britain as the world's premier economic power.

glaciers, rising sea levels, and global disaster, Matthew Boulton himself might have had second thoughts about building machines "for all the world." But probably not; John Ruskin might have lived just as well in a preindustrial world, but pretty much everyone else has done a whole lot better. From 1700 to 2000, the world's population has increased twelvefold—but its production of goods and services a hundredfold.

This is why, against all odds, the first decades of industrialization actually have something useful to say about their long-term impact on the world's climate—though it isn't what either side in the global warming debate would probably endorse. It certainly doesn't give much comfort to anyone who thinks humanity can be persuaded to spend any more for power than it has to; America and Europe might have finally so enriched themselves that they can afford to convert to wind, water, and solar power, but neither China nor India is likely to choose either over coal costing one-tenth as much. If the history of steam power teaches anything, it is that the lower-cost fuel option always wins. Right now, that option is about a trillion tons of easily mined, dirty, carbon-rich coal.

What this means, given the very real dangers of climate change, is that any comprehensive solution is going to have to do one of two things: figure out how to return all that carbon to where it was before humans learned how to exhaust it into the atmosphere—the technical term for putting carbon back is *sequestration*—or come up with a non-carbon-producing energy system that costs less than coal. Both options put the highest possible premium on invention. Phrased another way: There may be no way to put the genie of sustained invention back in the bottle, but we can put the genie to work.

By now, readers will have made up their minds about whether inventions and inventors deserve to hold center stage in the three-hundred-year dramatic crisis triggered by the genie's appearance. The previous three hundred or so pages have been largely an attempt to demonstrate how the inventions created the crisis, how the inventors created the inventions, and even how the birth of an idea about property "created" enough inventors to get the whole drama moving. The next few will try to examine what kept it moving, and moving in the same direction.

IF THERE IS ONE consistent theme in the story of innovation, it is its reflexive character. Without deep coal mines, there would not only have been no need for steam-powered pumps to drain them, there would have been no fuel for the pumps. The cast iron used to manufacture boilers, cylinders, pistons, and gears had impurities hammered from its "blooms" by steam-driven hammers. The primary cargo for the first coal-driven locomotives was coal itself; a close second was the iron ore that was smelted and wrought into six-foot rail segments. These are all examples of the capacity of technological advances to spill over into the economy at large, and so multiply their initial effects; Wilkinson's 1774 patent on his boring machine didn't just enrich the inventor, but enabled the growth of Boulton & Watt.

Technological spillovers aren't the only kind that matter economically. In a classic 1991 paper, the future Nobel Prize–winning economist Paul Krugman identified what he called "pecuniary spillovers": the tendency of industry to cluster in order to exploit lower costs, from both economies of scale and lowered transportation expense. Once a tipping point is reached—once an economy derives more value from making things than, for example, growing them—manufacturing will tend to establish itself in regions with other manufacturers, attracting still more manufacturers.

Krugman's economic geography is partly about space, explaining why some areas of the globe are wealthier than others. But it is also, even more importantly, about time. The term in general use is that economic growth is highly "path-dependent"—that is, once started down a path of growth, a society tends to continue on that path. As Krugman himself put it, "Small changes in the parameters of the economy may have large effects on its qualitative behavior . . . when some index that takes into account transportation cost, economies of scale, and the share of nonagricultural goods in expenditures crosses a critical threshold, population will start to concentrate and regions to diverge; *once started, this process will feed on itself*" (emphasis added).

For the last three centuries, the process has been feeding very well throughout the world. But the best feeding has been in places with a distinct judicial, political, and even linguistic history: what Winston Churchill called the English-speaking peoples, what jurists call the common law world, and what a number of modern scholars have termed the Anglosphere—Great Britain and its former colonies, including the United States, Canada, Australia, and New Zealand (though not colonies with more robust indigenous cultures, such as India, Hong Kong, South Africa, and so on).*

The last three centuries of the Anglosphere, whatever its current liabilities in a contemporary, multicultural world, are a reproach to the rather vaguely worded "Cardwell's Law," the creation of the economic historian D.S.L. Cardwell, who propounded it in 1972. Cardwell's Law contends that no nation maintains technological superiority for more than two or three generations, or, in Professor Cardwell's own words, "no nation has been very creative for more than an historically short period." The economic statistics tell a different story.

In 1700, when Great Britain's per capita gross domestic product was roughly equal to that of Italy, the aggregate world GDP was $371.3 billion in constant 1990 dollars. The Anglosphere's share, virtually all of it Britain, was a little more than 3 percent. By 1820, the conventional endpoint of the Industrial Revolution, the world's GDP had nearly doubled, to $694.5 billion, but the Anglosphere's share had grown even faster: to nearly $50 billion, or more than 7 percent.

At that moment, virtually all of the core inventions of industrialization had, in the economist Alfred Marshall's phrase,

* This is an argument starter, but not a particularly original one. Andrew Roberts argues in his controversial *History of the English Speaking Peoples Since 1900* that the distinction between the period of British ascendancy and of American is roughly the same as that between imperial Rome and its republican ancestor, i.e. not much of a difference at all. The Indian economic historian Deepak Lal believes the rise to predominance of the English-speaking world is the most important event of the last thousand years.

spilled over to the rest of Europe and much of the rest of the world. Moreover, by 1850, the Anglosphere's "secret"—an absolute and relative advantage in the tinkerers who specialized in the micro-inventions essential to constant improvement of complex machinery—had lost much of its value. The age of scientific invention was overtaking the age of intuitive invention, and France and Germany had started to benefit from it.* In 1850, France alone issued 2,272 patents, more than Britain and the U.S. combined.

However, fifty years later, in 1870, while the world economy had doubled again, to $1.11 trillion, the share of the Anglosphere was more than 19 percent. In 1900, by this time largely owing to the dramatic growth of the United States, just under 27 percent of the world economy was Anglophonic. In 1940, it was more than 30 percent, and in 1950—after the huge damage of the Second World War—it hit its all-time high of 37 percent; but even in 2000, it had "fallen" only to 28 percent of what was then a $36.7 trillion world economy.

Part of this was population growth. In 1700, the Anglosphere represented less than 2 percent of the world's population; by 1870, it was more than 6 percent, largely because of the decreased infant mortality and extended life spans directly traceable to increasing income. Since 1870, however, the share of world population has generally been between 6 and 8 percent—and its share of world income has been at least four times greater. The lead has been remarkably durable, even as compared to nations close to the Anglosphere in technical sophistication. Even as late as 2000, the Anglosphere's per capita GDP was still 36 percent greater than the Western European average—$26,238 vs. $19,264 (and 25 percent more than Japan's $21,051). Path dependence goes far to explain the durability of the Anglosphere's dominance of the world's economy for at least a century and a half after the spread

* Not without some envy; Goethe wrote that Germans "regard discovery and invention as a splendid personally gained possession, but the clever Englishman transforms it, by a patent, into a real possession. . . . One may well ask why are they in every respect in advance of us?"

of industrialization; in Paul Krugman's words, the phenomenon was feeding on itself. Economic prosperity resulted in more economic prosperity.

Which explains why the economic advantage persists, not how it began. It is impossible to look at the last three hundred years without wondering at the persistent advantage accrued during the first decades of the Industrial Revolution, and wondering at its cause.

Or, more to the point, wondering *who* caused it. In 1963, the historian of science Derek de Solla Price (who was, as it happens, the first to study the Antikythera mechanism in detail) wrote a book entitled *Little Science, Big Science,* in which he formulated the Price Law (building on the work of others, most especially the early twentieth-century economist Wilfredo Pareto). The Price Law states that the number of individuals responsible for half of all innovations is roughly equal to the square root of the number of total contributors.

Let's do the math. If the transformation of the world into an industrial economy depended on thirty thousand or so innovations, large and small (this is a generous number; by 1820, the United Kingdom and United States together had issued fewer than ten thousand patents), and the typical inventor was responsible for three of them—which probably understates the case, given the number generated by men like Watt, Roberts, Bramah, and Evans—then the total number of innovators is around ten thousand. By Price's Law, therefore, half of the Industrial Revolution is the work of about one hundred people. Which is a convenient number: one can far more easily visualize a pantheon of heroes with a hundred members than one with three thousand.

The figure of the "heroic" engineer has been sitting on a carnival dunking stool for more than a century, and one historian after another has taken a turn at knocking him into the water. Part of this is the natural tendency of scholars toward revisionism; if one generation of historians built godlike statues of Watt and Arkwright, then their successors were certain to find feet made of, if not clay, then at least base metal.

But there is more to the antiheroic school of industrial history than simply the Academy's demands for original scholarship. The unique characteristic of the Industrial Revolution—its sustainability—depended less on "macro-inventions" such as Watt's separate condenser than on the hundreds of micro-inventions that surrounded it: Henry Maudslay's leadscrew, Matthew Murray's D-valve, Richard Trevithick's fusible plug, and a thousand other improvements owed to a thousand other inventors. Their relative anonymity is an unavoidable consequence of the relative elevation of the Watts and the Arkwrights.

This is why Samuel Smiles, the great hagiographer of British inventions, was in some ways as important as his subjects. It was Smiles, with his biographies of Boulton, Watt, Stephenson, and others, who made a secular religion out of human striving: "It is not the man of the greatest natural vigour and capacity who achieves the highest results, but he who employs his powers with the greatest industry and the most carefully disciplined skill. . . ."

It matters less that Smiles was right than that he was believed. It was belief in the chance to become a national hero that, despite equal distribution of talent, and even of expertise, at least throughout Europe, made innovation such a local phenomenon. It is suggestive that Anders Ericsson's "expert performance" calculations reinforce this: A typical apprenticeship, running as it did for seven years, or fifteen to twenty thousand hours, buys an expert weaver, goldsmith, or millwright; that's what the apprenticeship system was supposed to do. But the commitment to spend a similar amount learning to be an expert *inventor* is a lousy economic decision for almost everyone who makes it. Not only does it mean years of working without income—forgoing the income that could have been generated by working at the skill acquired while an apprentice—but it offers no guarantee that the investment will be recouped. While one should be cautious about applying current data to earlier historical eras, it's worth recalling that twentieth-century inventors, on average, sacrificed nearly one-third of their potential lifetime income.

Thus, despite the fact that those hours, ten thousand at a time, were being invested in skills acquisition all over Europe, the only place where really large numbers of skilled craftsmen decided to become inventors was in the Anglosphere. And it wasn't that they made enough money to make it a smart decision, so much as the fact that they *thought* they could.

Nowhere did that thought take root more firmly than in the part of the Anglosphere known as the United States of America.

ON THE THIRD FLOOR of the National Museum of American History, part of the Smithsonian Institution in Washington, D.C., is a scale model, constructed of varnished wood, that looks a little like an outrigger canoe with sixteen vertical tubes inserted fore and aft. On May 22, 1849, the device, "a new and improved manner of combining adjustable buoyant air chambers with a steam boat or other vessel for the purpose of enabling their draught of water to be readily lessened to enable them to pass over bars, or through shallow water," received U.S. patent number 6469. Its inventor was "Abraham Lincoln, of Springfield, in the County of Sangamon, in the State of Illinois."

Lincoln, the only American president ever awarded a patent, had a long and passionate love for things mechanical. He made his living for many years as a railroad lawyer and appears to have absorbed something of the fascination with machines, and with steam, of the engineers with whom he worked. In his first public speech, in 1832, he spent an inordinate amount of time talking about the need for navigable rivers and canals to accommodate steamboats. In the middle of the Civil War, he signed, on July 1, 1862, the Pacific Railway Act, the authorizing legislation for what would become America's transcontinental railroad. Even more revealing, in 1859, after his loss in the Illinois senatorial race against Stephen Douglas, he was much in demand for a speech entitled "Discoveries, Inventions, and Improvements" that he gave at agricultural fairs, schools, and self-improvement societies.

The speech—decidedly not one of Lincoln's best—nonetheless revealed an enthusiasm for mechanical innovation that resonates powerfully even today. "Man," Lincoln said, "is not the only animal who labors, but he is the only one who *improves* his workmanship . . . by *Discoveries* and *Inventions*."

The speech goes on to offer a brief history of civilization as seen through the lens of one invention after another:

> We can scarcely conceive the possibility of making much of anything else, without the use of iron tools. . . . the boat is indispensable to navigation [though] it is not probable that the philosophical principle upon which the use of the boat primarily depends—to-wit, the *principle* that anything will float, which cannot sink without displacing more than its own weight of water—was known . . . the plow, of very early origin; and reaping and threshing machines, of modern invention, are, at this day, the principal improvements in agriculture. . . . Take any given space of the earth's surface—for instance, Illinois; and all the power exerted by all the men, and beasts, and running water, and steam, over and upon it, shall not equal the one hundredth part of what is exerted by the blowing of the wind over and upon the same space. And yet no very successful mode of controlling, and directing the wind, has been discovered. . . .

Lincoln concluded:

> The advantageous use of *Steam-power* is, unquestionably, a modern discovery. And yet, as much as two thousand years ago the power of steam was not only observed, but an ingenious toy was actually made and put in motion by it, at Alexandria in Egypt. What appears strange is that neither the inventor of the toy, nor any one else, for so long a time afterwards, should perceive that steam would move *useful* machinery as well as a toy. . . . in the days before Edward Coke's original Statute on Monopolies, any man could instantly use what another had invented; so that the inventor had no special advantage from his own invention. . . . The

> patent system changed this; secured to the inventor, for a limited time, the exclusive use of his invention; and thereby added the *fuel* of interest to the *fire* of genius, in the discovery of new and useful things.

There's something appealing about starting a book in a museum in London and ending it in one in Washington. However, the real ending takes place just to the north of the National Museum of American History, at another Washington landmark: the United States Department of Commerce, a huge pile of limestone known since 1983 as the Herbert C. Hoover Building. Since 1995, it has housed the White House Visitor Center on its street level, in the same space that was once home to the United States Patent and Trademark Office.

Incised in the stone over the Herbert C. Hoover Building's north entrance is the legend that, with Lincoln's characteristic brevity, sums up the single most powerful idea in the world:

THE PATENT SYSTEM ADDED
THE FUEL OF INTEREST
TO THE FIRE OF GENIUS

ACKNOWLEDGMENTS

Writing may be a lonely business, but it is not a solitary one. Like the steam engine itself, creating *The Most Powerful Idea in the World* depended on the work of hundreds if not thousands of people, but—as with the steam engine—only a few can be adequately recognized. Life isn't fair.

We are not yet at the moment in history where research can be performed entirely via Internet access, and I am small-*l* luddite enough to be grateful for the need to spend time in libraries, large and small. As always, the guidance of Elizabeth Bennett at Princeton University's Firestone Library has been indispensable; so indeed has the Princeton Public Library, and I am delighted to recognize the help of Leslie Burger and her staff. I am also profoundly in debt to the Birmingham Central Library, which has maintained—and microfilmed—the Boulton and Watt Archive, and the Matthew Boulton Papers. A special debt of gratitude is similarly owed to Tom Vine, at London's Science and Society Picture Library.

I am likewise happy to recognize the enormous help of two people who signed on to help me to enforce quality-control standards on specific chapters of the manuscript for *The Most Powerful Idea:* Dan Swain and Paul Anderson. Each is responsible for saving me from hundreds of embarrassing errors, and neither can be blamed for any that crept in after their work was done. As both were found through the good offices of Princeton University, my advice to all authors remains: Live in a college town.

The subtitle to this book reads "*A* Story of Steam, Industry, and

Invention," and the use of the indefinite article is no accident. The story of humanity's climb out of its Malthusian trap has been told from a hundred different perspectives, and all of them have something useful to say. The scholars, living and dead, who have traveled this road before me are recognized in the endnotes that follow these Acknowledgments, but the work of nine didn't supply merely historical examples but inspiration: Joel Mokyr, David Landes, Gregory Clark, Eugene Ferguson, John Lienhard, Lynn White, Samuel Florman, David Warsh, and—most of all—Abbott Payson Usher. Friends and colleagues who read this book in various draft forms have given service above and beyond the call of duty: John Rosen, Michel Debiche, Frank Ryle, Joyce Howe, Holly Goldberg Sloan, Gary Rosen, and David Jacobus. While I was writing this book, Lewis Lapham and Adam Garfinkle were both kind enough to offer me the chance to write for their respective publications, for which I am happy to publicly thank them.

Thanks are due also to this book's editors. At Jonathan Cape, Ellah Allfrey was this book's first champion, and her editing added immeasurably to it before she passed the baton to Alex Bowler, who has gracefully seen it through to publication. Tim Bartlett at Random House immersed himself in the manuscript of *The Most Powerful Idea* with a level of detail and care that I have never seen surpassed—and as a onetime editor myself, this is no small feat. There are literally thousands of places in this text that have his imprint. I promised Tim that I would immunize him from any criticisms of this book's editing, and do so cheerfully: Its many remaining infelicities appear, without fail, in areas where I disagreed with Tim's extraordinarily perceptive edits. Thanks are also due to Tim's unfailingly helpful assistant, Jessie Waters. The book's production editor, Janet Wygal, and its copy editor, Emily DeHuff, delivered work at the very highest level of professionalism, and *The Most Powerful Idea* is better for their labors. A special thank-you is owed to Jennifer Hershey, who has consistently shown her support for this project.

My agents, Eric Simonoff of William Morris Endeavor, and Anne Sibbald at Janklow & Nesbit, have never done anything but surpass the highest expectations of what a literary agent can be: immediate response to questions, intelligent evaluations of work, determined promotions of the interests of author *and* book. It is one of the joys of my life to count them as friends and advocates.

Most important, I am grateful to my wife, Jeanine, and to my children: Quillan (one of the book's very first readers), Emma, and Alex. I am no treat to live with, even when I am not writing a book, and their love and support is the only thing that makes the life of a writer even possible.

NOTES

PROLOGUE: *ROCKET*

xv **"the rapid development in industry"** George N. Clark, *The Idea of the Industrial Revolution* (Glasgow: Jackson, Son & Company, 1953).

xv **Carolingian merchants spoke different languages** This has been conclusively demonstrated by dozens of studies, of which the most recent is Gregory Clark, *A Farewell to Alms: A Brief Economic History of the World* (Princeton: Princeton University Press, 2007).

xvi **The worldwide per capita GDP in 800 BCE** Michael Kremer, "Population Growth and Technological Change: One Million B.C. to 1990," *Quarterly Journal of Economics* 108, no. 3, Fall 1993. The figures in question are J. Bradford deLong's slightly different estimates.

xvii **The nineteenth-century French infant** Numbers from UN and CIA Factbook.

xviii **A skilled fourth-century weaver** Kirkpatrick Sale, *Rebels Against the Future: The Luddites and Their War on the Industrial Revolution—Lessons for the Computer Age* (Reading, MA: Addison-Wesley, 1995).

xviii **But by 1900** In 1900, the average U.S. hourly wage was $0.22, and a loaf of bread cost about a nickel; in 2000, the average wage was $18.65, and a loaf of bread cost less than $1.79.

xxi **"[a]bout 1760, a wave of gadgets swept over England"** T. S. Ashton, *Industrial Revolution* (Oxford: Oxford University Press, 1997).

xxii **"fizzled out"** Joel Mokyr, "The Great Synergy: The European Enlightenment as a Factor in Modern Economic Growth," April 2005, online article at http://faculty.weas.northwestern.edu/~jmokyr/Dolfsma.pdf.

CHAPTER ONE: CHANGES IN THE ATMOSPHERE

5 **No other steam engines were inspired by it** T. P. Tassios, "Why the First Industrial Revolution Did Not Take Place in Alexandria," in *10th International Symposium on Electrets, 1999*, IEEE, eds. (Athens: IEEE, 2002).

6 **"if a light vessel with a narrow mouth"** Hero of Alexandria, Joseph

George Greenwood, and Bennett Woodcroft, *The Pneumatics of Hero of Alexandria* (London and New York: Macdonald and American Elsevier, 1971).

7 **"very satisfactory theory"** Marie Boas, "Hero's Pneumatica: A Study of Its Transmission and Influence," in Otto Mayr, ed., *Philosophers and Machines* (New York: Neale Watson Academic Publications, 1976).

8 **Aleotti's work, and subsequent translations** Ibid.

8 **It is testimony to the weight of formal logic** Graham Hollister-Short, "The Formation of Knowledge Concerning Atmospheric Pressure and Steam Power in Europe from Aleotti (1589) to Papin (1690)," *History of Technology* 25, 2004.

8 **"What is so intricate"** Robert Burton, *Anatomy of Melancholy,* cited in Boas, "Hero's Pneumatica."

9 **Torricelli had not only invented** W. E. Knowles Middleton, "The Place of Torricelli in the History of the Barometer," *Isis: Journal of the History of Science in Society* 54, no. 1, March 1963.

11 **"technical wonders of its time"** Lynn White, *Medieval Technology and Social Change* (London: Oxford University Press, 1964).

12 **As the air was pumped out of the chamber** Arnold Pacey, *The Maze of Ingenuity: Ideas and Idealism in the Development of Technology* (London: Lane, 1974).

13 **A dispute between King and Parliament** H.C.G. Matthew and B. Harrison, eds., *Oxford Dictionary of National Biography: In Association with the British Academy: From the Earliest Times to the Year 2000* (Oxford and New York: Oxford University Press, 2004).

14 **"the leading pumping engineer in England"** Allan Chapman, "England's Leonardo: Robert Hooke and the Art of Experiment in Restoration England," *Proceedings of the Royal Institution of Great Britain* 67, 1996.

14 **to anything, in short** Ibid.

15 **"those two grand and most catholic principles, matter and motion"** "Robert Boyle" in Noretta Koertge, ed., *New Dictionary of Scientific Biography* (Detroit: Scribner's, 2008).

15 **halfway to atheism** "Robert Boyle" in *Oxford Dictionary of National Biography.*

15 **"engine philosophy"** Steven Shapin, Simon Schaffer, and Thomas Hobbes, *Leviathan and the Air-Pump: Hobbes, Boyle, and the Experimental Life, Including a Translation of Thomas Hobbes,* Dialogus physicus de natura aeris *by Simon Schaffer* (Princeton, NJ: Princeton University Press, 1985).

16 **England's most gifted mathematician** Chapman, "England's Leonardo."

16 **"the best Mechanick this day in the world"** Ibid.

16 **"*Gentleman,* free, and unconfin'd"** Ibid.

16 **made him the first scientist in British history** Ibid.

16 **It took until 1665** "Robert Hooke" in *Oxford Dictionary of National Biography.*

17 **"mere Empiricks"** Shapin, Schaffer, and Hobbes, *Leviathan and the Air-Pump.*

CHAPTER TWO: A GREAT COMPANY OF MEN

18 **In 1671, he got the chance** Cornelis D. Andriesse, *Huygens: The Man Behind the Principle* (Cambridge, UK, and New York: Cambridge University Press, 2005).

20 **In the 1686 issue of *Philosophical Transactions*** White, *Medieval Technology and Social Change.*

20 **"Since it is a property of water"** Milton Kerker, "Science and the Steam Engine," *Technology and Culture* 2, no. 4, Autumn 1961.

21 **By the time he built a demonstration submarine** Richard S. Westfall, *The Galileo Project* (Rice University), at http://galileo.rice.edu/.

23 **"to Raise Water from Lowe Pitts by Fire"** Pacey, *Maze of Ingenuity.*

24 **"was responsible for the design and fabrication"** Anthony F. C. Wallace, *The Social Context of Innovation: Bureaucrats, Families, and Heroes in the Early Industrial Revolution, As Foreseen in Bacon's* New Atlantis (Princeton, NJ: Princeton University Press, 1982).

24 **"a place of resort for artists, mechanics"** A 1640 letter to Robert Boyle, cited in ibid.

25 **"at a potter's house in Lambeth"** Ibid.

27 **to refill the boiler at least once a minute** Richard L. Hills, *Power from Steam* (Cambridge: Cambridge University Press, 1989).

27 **"Mr. Savery . . . entertained the Royal Society"** *Philosophical Transactions of the Royal Society* XXI, p. 228.

28 **"one of the great original synthetic inventions"** Eugene S. Ferguson, "The Steam Engine before 1830" in Kranzberg and Pursell, eds., *Technology in Western Civilization* (New York: Oxford University Press, 1967).

29 **The 1712 engine of Thomas Newcomen** Howard Jones, *Steam Engines: An International History* (London: Ernest Benn Limited, 1973).

30 **(sometimes a plumber)** David Richards, "Thomas Newcomen and the Environment of Innovation," *Industrial Archaeology* 13, no. 4, Winter 1978.

31 **of the latter's progress** Kerker, "Science and the Steam Engine."

31 **"could he [Papin] make a speedy vacuum"** Abbott Payson Usher, *A History of Mechanical Inventions* (Cambridge, MA: Harvard University Press, 1954); also Samuel Smiles, *Men of Invention and Industry* (New York: Harper & Brothers, 1885). The story of the Newcomen-Hooke correspondence dates from 1797, when it was reported by Dr. John Robison, a friend of James Watt, whose other recollections have almost uniformly proven valid. In the absence of corroborating documentary evidence, however, some scholars accept it, others not—though much of the doubt seems to come from the belief that the fifty-two-year-old Hooke would have had little to say to a twenty-four-year-old ironmonger; in short, retroactive snobbishness.

32 **For purposes of the experiment** Martin Triewald's 1734 *Short Description of the Atmospheric Steam Engine,* quoted in Hills, *Power from Steam.*

34 **"not being either philosophers"** "Thomas Newcomen" in *Oxford Dictionary of National Biography.*

34 **Newcomen and Calley replaced** Usher, *History of Mechanical Inventions.*

34 **"the Air makes a Noise"** Hills, *Power from Steam.*

35 **"the valve still functioned perfectly"** Ibid.

35 **They also tell of the years he spent** Sir William Fairbairn, *The Life of Sir William Fairbairn* (London, 1877), quoted in Cohen.

36 **Now imagine producing such a fitting** Joseph W. Roe, *English and American Tool Builders* (New Haven: Yale University Press, 1916).

36 **"partly because they were equipped"** David Wolman, *A Left-hand Turn Around the World: Chasing the Mystery and Meaning of All Things Southpaw* (Cambridge, MA: Da Capo Press, 2005).

36 **The hand has led the brain to evolve** Richard Sennett, *The Craftsman* (New Haven: Yale University Press, 2008).

38 **"It is only with Leonardo"** Usher, *History of Mechanical Inventions.*

38 **"Pyramids, cathedrals, and rockets"** E. S. Ferguson, "The Mind's Eye: Nonverbal Thought in Technology," *Science* 197, no. 4306, August 1977.

39 **"both dead and living"** Usher, *History of Mechanical Inventions.*

39 **"During the up-stroke"** Hills, *Power from Steam.*

40 **"the greatest single act of synthesis"** Usher, *History of Mechanical Inventions.*

40 **"designing to turn his engines"** *Newcommen* (sic) *v. Harding,* TNA: PRO, C11/1247/38, 39, at http://www.nationalarchives.gov.uk

41 **"divided the profit to arise"** "Thomas Savery" in *Oxford Dictionary of National Biography.*

41 **Though Newcomen's take** Eric Roll of Ipsden, *An Early Experiment in Industrial Organisation, Being a History of the Firm of Boulton & Watt, 1775–1805* (London and New York: Longmans, Green, 1930).

41 **"Whereas the invention for raising water"** "Thomas Newcomen" in ibid.

CHAPTER THREE: THE FIRST AND TRUE INVENTOR

44 **Though already in possession of an income** The estimate was made by Thomas Wilson, the Keeper of Records for the Office of His Majesty's Papers and Records; Catherine Drinker Bowen, *The Lion and the Throne: The Life and Times of Sir Edward Coke (1552–1634)* (Boston: Little, Brown, 1957).

45 **one estimate puts Coke's take at £100,000** "Edward Coke" in *Oxford Dictionary of National Biography.*

45 **"country gentlemen, acquisitive parsons"** Ibid.

47 **The idea of exclusive commercial franchises** Kenneth W. Dobyns, *History of the United States Patent Office: The Patent Office Pony* (Fredericksburg, VA: Kirkland Museum, 1994).

47 **the emperor had the unfortunate soul executed** Ibid.

48 **Coke was convinced that monopolies were costly** In *Davenant v. Hurdis*, Coke, again as Attorney General, argued against another potential monopoly, this one a restriction by the powerful guild known as the Merchant Tailors of London. The tailors required that their members use other members on at least half the cloth they cut, effectively (or so argued Coke) creating a monopoly. See Samuel E. Thorne, *Sir Edward Coke, 1552–1952* (London: Quaritch, 1957).

49 **decades before *Darcy v. Allein*** Barbara Malament, "The 'Economic Liberalism' of Sir Edward Coke," *Yale Law Journal* 76, no. 7, June 1967.

49 **In order to find a precedent** Thorne, *Sir Edward Coke.*

49 **"the franchises and privileges"** D.O. Wagner, "Coke and the Rise of Economic Liberalism," *Economic History Review* 6, no. 1, October 1935.

50 **the Netherlands' States-General** Mario Biagioli, "Early Modern Instruments Database: An Appendix to *From Prints to Patents: Living on Instruments in Early Modern Europe,*" *History of Science* 44, 2006.

51 **"generally inconvenient"** Vishwas Devaiah, "A History of Patent Law," 2006, online article at http://www.altlawforum.org/PUBLICATIONS/document.2004-12-18.0853561257.

54 **And he liked it** "Francis Bacon" in *Oxford Dictionary of National Biography.*

55 **to cite it in *Darcy v. Allein*** D. O. Wagner, "The Common Law and Free Enterprise: An Early Case of Monopoly, *Economic History Review* 7, no. 2, May 1937.

56 **"men of far greater titles"** Bowen, *The Lion and the Throne: The Life and Times of Sir Edward Coke (1552–1634).*

56 **"we shall never see his like again"** Thorne, *Sir Edward Coke.*

56 **"Mr. Attorney: I respect you"** "Francis Bacon" in *Oxford Dictionary of National Biography.*

58 **"college for Inventors"** Wallace, *Social Context of Innovation.*

58 **Tellingly, though Bacon had respect** Ibid.

58 **"the inventor of ordnance and of gunpowder"** Francis Bacon, *The Advancement of Learning and New Atlantis* (London: Oxford University Press, 1956).

59 **Bacon's faith in progress** Wallace, *Social Context of Innovation.*

60 **"had so small satisfaction from his studies"** "John Locke" in *Oxford Dictionary of National Biography.*

61 **in his four decades as a member of the RS** Ibid.

63 **"disposing of the affairs of the kingdom"** Henry Ireton, quoted in E. P. Thompson, *The Making of the English Working Class* (Harmondsworth: Penguin, 1968).

63 **they excluded servants and beggars** Crawford B. Macpherson, *The Political Theory of Possessive Individualism: Hobbes to Locke* (Oxford: Clarendon Press, 1962).

64 **"You and your ancestors got your propriety"** Micheline Ishay, ed., *The*

Human Rights Reader: Major Political Writings, Essays, Speeches, and Documents from the Bible to the Present (New York: Routledge, 1997).

64 **"Let no young wit be crushed"** G. N. Clark, "Early Capitalism and Invention," *Economic History Review* 6, no. 2, April 1936.

65 **"Nature furnishes us only with the material"** Peter King, *The Life and Letters of John Locke, with Extracts from His Journals and Commonplace Books: With a General Index* (New York: B. Franklin, 1972).

65 **It is scarcely surprising** Nigel Stirk, "Intellectual Property and the Role of Manufacturers: Definitions from the Late Eighteenth Century," *Journal of Historical Geography* 27, no. 4, 2001, citing Jenny Uglow, *Hogarth: A Life and a World* (London: Faber and Faber, 1997) (who also cites Lord Chesterfield's famous "Wit, my lords, is a sort of property").

66 **"nonsense on stilts"** Paul E. Sigmund, *The Selected Political Writings of John Locke* (New York: W.W. Norton, 2005).

66 **"God . . . made the right of work"** Arnold Toynbee and Benjamin Jowett, *Lectures on the Industrial Revolution of the 18th Century in England: Popular Addresses, Notes and Other Fragments* (London and New York: Longmans, Green, 1902).

CHAPTER FOUR: A VERY GREAT QUANTITY OF HEAT

67 **"the steam engine has done much more for science"** Though this is traditionally attributed to Kelvin, a better (though still unreliably) documented author is the early twentieth-century Harvard physiologist and chemist Lawrence Joseph Henderson. David Philip Miller, "Seeing the Chemical Steam Through the Historical Fog," *Annals of Science* 65, no. 1, January 2008.

68 **"Aristotle asserts that cabbages produce caterpillars"** Martin Goldstein and Inge F. Goldstein, *The Experience of Science: An Interdisciplinary Approach* (New York: Plenum Press, 1984).

71 **some acceptable sinecure** Mokyr, "The Great Synergy."

72 **"the buzzword of the eighteenth century"** Ibid.

72 **J. T. Desaguliers, the same critic** Ibid.

73 **"the Riches, Honour, Strength"** Ibid.

74 **Two years before his death in 1704** Robert Horwitz and Judith Finn, "Locke's Aesop's Fables," *The Locke Newsletter* no. 6, Summer 1975.

75 **The first was the notion that heat** D.S.L. Cardwell, *From Watt to Clausius: The Rise of Thermodynamics in the Early Industrial Age* (London: Heinemann Educational, 1971).

79 **"fixed air"** Henry Marshall Leicester and Herbert Klickstein, *A Source Book in Chemistry, 1400–1900* (Cambridge, MA: Harvard University Press, 1963).

79 **"a violence equal to that of gunpowder"** W. F. Magie, *A Source Book in Physics* (Cambridge, MA: Harvard University Press, 1963).

79 **He then placed the water over heat** Cardwell, *From Watt to Clausius.*

80 **"I, therefore, set seriously about making experiments"** Magie, *Source Book in Physics.*

80 **He heated a pound of gold** Cardwell, *From Watt to Clausius.*

82 **"the first great achievement"** White, *Medieval Technology and Social Change.*

82 **They were, for example, common in northern Europe** Ibid.

84 **Europe's first true "wood crisis"** Ibid.

84 **It also meant a lot more wood** John H. Lienhard, *How Invention Begins: Echoes of Old Voices in the Rise of New Machines* (Oxford and New York: Oxford University Press, 2006).

85 **by the early fifteenth century** Barbara Freese, *Coal: A Human History* (New York: Perseus Books, 2003).

86 **It was not until the 1600s** Lienhard, *How Invention Begins.*

88 **"in a state within a state"** Margaret T. Hodgen, *Change and History* (New York: Wenner-Green Foundation for Anthropological Research, 1952).

88 **15 percent of the total were for drainage alone** Wallace, *Social Context of Innovation.*

88 **In 1752, a study was made** Pacey, *Maze of Ingenuity.*

89 **as late as the 1840s** Joel Mokyr, *The Lever of Riches: Technological Creativity and Economic Progress* (New York: Oxford University Press, 1990).

CHAPTER FIVE: SCIENCE IN HIS HANDS

90 **In 1747, one of them** Joseph Irving, *The Book of Dumbartonshire* (Edinburgh and London: W. and A. K. Johnston, 1879).

91 **"to clean them and to put them in the best order"** Glasgow University press office, 1998.

91 **during the eighteenth century** Alexander Broadie, *The Cambridge Companion to the Scottish Enlightenment* (Cambridge and New York: Cambridge University Press, 2003).

92 **"the first six, with the eleventh and twelfth Books"** Ibid.

93 **"it is the principal sustenance"** Henry Fielding, "An Inquiry into the Late Increase in Robbers," in Ronald Paulson, *Henry Fielding: The Critical Heritage* (London: Routledge, 1995).

94 **it had been founded "only" in 1631** Thomas H. Marshall, *James Watt (1736–1819)* (London and Boston: L. Parsons and Small, 1925).

94 **"foreigners, alien or English"** Ibid.

94 **On the other hand, his willingness to leave London** Ibid.

95 **"to work as well as most journeymen"** "James Watt," in *Oxford Dictionary of National Biography.*

95 **"large, stately, and well-built city"** George MacGregor, *The History of Glasgow: From the Earliest History to the Present Time* (London: Hamilton & Adams, 1881).

96 **"every thing became Science"** Eric Robinson and A. E. Musson, *James Watt and the Steam Revolution: A Documentary History* (London: Adams & Dart, 1969).

96 **"Allow me to give an instance"** Ibid.

98 **"set about repairing it"** Birmingham Central Library (Birmingham, England) and Adam Matthew Publications, *The Industrial Revolution: A Documentary History. Series One: The Boulton and Watt Archive and the Matthew Boulton Papers from the Birmingham Central Library* (Marlborough, Wiltshire, England: Adam Matthew Publications, 1993).

98 **"the toy cylinder exposed a greater surface"** Marshall, *James Watt.*

99 **Most textbooks plot a "boiling curve"** Hasok Chang, *Inventing Temperature: Measurement and Scientific Progress* (Oxford and New York: Oxford University Press, 2004).

100 **two different "boyling" temperatures** Ibid.

100 **The measurement problem was acute enough** Ibid.

100 **One small example of it** Mokyr, "The Great Synergy."

100 **In one of his notebooks** Richard L. Hills, "The Origins of James Watt's Perfect Engine," *Transactions of the Newcomen Society* 68, 1997.

101 **"I mentioned it to my friend Dr. Black"** Donald Fleming, "Latent Heat and the Invention of the Watt Engine," in Mayr, ed., *Philosophers and Machines.*

101 **Watt didn't discover the existence of latent heat** Ibid.

102 **Heating the cylinder walls** Hills, "The Origins of James Watt's Perfect Engine."

103 **"ran on making engines *cheap*"** James Patrick Muirhead, *The Life of James Watt, with Selections from His Correspondence* (London: J. Murray, 1858).

104 **"steam was an elastic body"** Birmingham Central Library (Birmingham, England) and Adam Matthew Publications, *The Industrial Revolution: A Documentary History. Series Three: The Papers of James Watt and His Family Formerly Held at Doldowlod House* (Marlborough, England: A. Matthew, 1998).

106 **"nearly as perfect"** F. M. Scherer, "Invention and Innovation in the Watt-Boulton Steam Engine Venture," in Kranzberg, ed., *Technology and Culture: An Anthology* (New York: Schocken Books, 1972).

106 **"I can think of nothing else"** Watt to Lind, April 29, 1765, in Robinson and Musson, *James Watt and the Steam Revolution.*

106 **"the invention was complete"** Scherer, "Invention and Innovation in the Watt-Boulton Steam Engine Venture."

107 **"A Company for carrying on an undertaking"** Charles Mackay, Josef Penso de la Vega, and Martin S. Fridson, *Extraordinary Popular Delusions and the Madness of Crowds* (New York: Wiley, 1996).

108 **"I am going on with the Modell"** Watt to Roebuck, September 9, 1765, in Robinson and Musson, *James Watt and the Steam Revolution.*

109 **As a result, he tried dozens of combinations** Scherer, "Invention and Innovation in the Watt-Boulton Steam Engine Venture."

109 **"Cotton was proposed"** Birmingham Central Library and Adam Matthew Publications, *The Industrial Revolution: A Documentary History. Series One: The Boulton and Watt Archive and the Matthew Boulton Papers from the Birmingham Central Library.*

109 **"Dear Jim . . . Let me suggest a method"** Ibid.

110 **"what I knew about the steam engine"** Ibid.

110 **"my principal hindrance"** Muirhead, *Life of James Watt.*

110 **"relief amidst [his] vexations"** Birmingham Central Library and Adam Matthew Publications, *The Industrial Revolution: A Documentary History. Series One: The Boulton and Watt Archive and the Matthew Boulton Papers from the Birmingham Central Library.*

111 **"have given me health and spirits"** Marshall, *James Watt.*

112 **"the Most compleat Manufacturer"** Jenny Uglow, *The Lunar Men: Five Friends Whose Curiosity Changed the World* (New York: Farrar, Straus and Giroux, 2002).

112 **"I would rather face a loaded cannon"** Scherer, "Invention and Innovation in the Watt-Boulton Steam Engine Venture."

113 **"I was excited by two motivs"** Boulton to Watt, February 7, 1769, in Robinson and Musson, *James Watt and the Steam Revolution.*

CHAPTER SIX: THE WHOLE THING WAS ARRANGED IN MY MIND

115 **"It was *in the Green of Glasgow*"** Robert Hart, "Reminiscences of James Watt," *Transactions of the Glasgow Archaeological Society* 1, no. 1, with commentary by John W. Stephens, at http://www.history.rochester.edu/steam/.

117 **By the 1990s, Ericsson's research was demonstrating** K. Anders Ericsson, "Creative Expertise as Superior Reproducible Performance: Innovative and Flexible Aspects of Expert Performance," *Psychological Inquiry* 10, no. 4, 1999.

120 **When a single neuron chemically fires** David Robson, "Disorderly Genius: How Chaos Drives the Brain," *New Scientist,* June 29, 2009.

121 **This was expected** M. Jung-Beeman, "Neural Activity When People Solve Verbal Problems with Insight," *PLoS Biology* 2, no. 4, April 2004.

121 **"The relaxation phase is crucial"** Jonah Lehrer, "The Eureka Hunt," *The New Yorker,* July 28, 2008.

121 **Some of the results were predictable** Joseph Rossman, *The Psychology of the Inventor: A Study of the Patentee* (Washington, D.C.: Inventors Publishing, 1931).

122 **"lack of capital"** Ibid.

123 **more than half will continue to invest their time** Thomas Astebro, "Inventor Perseverance After Being Told to Quit: The Role of Cognitive Biases," *Journal of Behavioral Decision Making* 20, January 2007.

123 **"*may* be inventors"** Scherer, "Invention and Innovation in the Watt-

Boulton Steam Engine Venture," citing Joseph Schumpeter's *Theory of Economic Development*.

123 **Another study, this one conducted in 1962** Donald W. MacKinnon, "Intellect and Motive in Scientific Inventors: Implications for Supply," in Simon Kuznets, ed., *The Rate and Direction of Inventive Activity: Economic and Social Factors* (Princeton: Princeton University Press, 1962).

124 **the eighteenth-century Swiss mathematician Daniel Bernoulli** Peter L. Bernstein, *Against the Gods: The Remarkable Story of Risk* (New York: John Wiley & Sons, 1996).

124 **"The more inventive an independent inventor is"** MacKinnon, "Intellect and Motive in Scientific Inventors: Implications for Supply," in Kuznets, ed., *Rate and Direction of Inventive Activity*.

125 **"first scientific man to study the Newcomen engine"** "Henry Beighton" in *Oxford Dictionary of National Biography*.

125 **Leonhard Euler applied** Usher, *History of Mechanical Inventions*.

128 **His published table of results** Jennifer Karns Alexander, *The Mantra of Efficiency: From Waterwheel to Social Control* (Baltimore: Johns Hopkins University Press, 2008).

128 **The resulting experiment** Pacey, *Maze of Ingenuity*.

128 **His example showed a generation of other engineers** Mokyr, "The Great Synergy," quoting Cardwell, 1994.

128 **"In comparing different experiments"** Pacey, *Maze of Ingenuity*.

129 **As far back as the 1960s** Dean Keith Simonton, "Creativity as Blind Variation and Selective Retention: Is the Creative Process Darwinian?" *Psychological Inquiry* 10, no. 4, 1999.

130 **"ideational mutations"** Ibid.

130 **"self-perpetuating feedback loops"** James Flynn, *What Is Intelligence? Beyond the Flynn Effect* (Cambridge and New York: Cambridge University Press, 2007).

131 **at some point, the recruits are going to reduce** Fritz Machlup, "The Supply of Inventors and Inventions," in Kuznets, ed., *Rate and Direction of Inventive Activity*.

132 **In Machlup's exercise** Ibid.

132 **"a statement that five hours of Mr. Doakes' time"** Ibid.

133 **"as the greatest and most useful man"** Hart, "Reminiscences of James Watt."

CHAPTER SEVEN: MASTER OF THEM ALL

136 **"woolen cowl for winter"** From *The Holy Rule of St. Benedict*, translated by Rev. Boniface Verheyen, OSB (Atchison, KS: Abbey Student Press, 1949).

136 **The monks of St. Victor's Abbey** Mokyr, *Lever of Riches*.

136 **The distinction was the work** Pauline Matarasso, *The Cistercian World:*

Monastic Writings of the Twelfth Century (London and New York: Penguin, 1993).

137 **In 1997, a team of archaeologists** R. W. Vernon, "The Geophysical Evaluation of an Iron-Working Complex: Rievaulx and Environs, North Yorkshire," *Archaeological Prospection* 5, no. 4, April 1998.

137 **from the random magnetism** Gerry McDonnell, "Geophysical Techniques Applied to Early Metalworking Sites," *The Historical Metallurgy Society*, Data Sheet #4, April 1995.

139 **Men were smelting iron in Coalbrookdale** The oldest surviving furnace at the site has a lintel carrying several dates, of which the earliest is 1638, but a dozen different operators produced iron at Coalbrookdale for decades both before and after its installation. Arthur Raistrick, *Dynasty of Iron Founders: The Darbys and Coalbrookdale* (London and New York: Longmans, 1953).

142 **"A certain quantity of iron ore"** Agricola, Herbert Hoover, and Lou Hoover, *Georgius Agricola* De re metallica (London: Mining Magazine, 1912). The Hoovers—the future president and his wife—make a good case that Agricola's description of iron manufacture was lifted, more or less unchanged, from a prior work by Vanoccio Biringuccio.

146 **The Dutch secret turned out to be** Usher, *History of Mechanical Inventions.*

146 **"a new way of casting iron bellied pots"** Samuel Smiles, *Industrial Biography: Iron-Workers and Tool-Makers* (Boston: Ticknor and Fields, 1864).

147 **"to make iron, steel, or lead"** List and Index Society, *Calendar of Patent Rolls* 30, Eliz. I, p. 57 (Kew, UK: National Archives, 2008).

147 **"sole priviledge to make iron"** "Simon Sturtevant" in *Oxford Dictionary of National Biography.*

147 **"the mystery, art, way, and means"** Thomas Webster, *Reports and Notes of Cases on Letters Patent for Inventions (1607–1855)* (London: Blenkarn, 1844).

147 **In the event, the younger Dudley** Gerald Newman and Leslie Ellen Brown, *Britain in the Hanoverian Age, 1714–1837: An Encyclopedia* (New York: Garland, 1997).

148 **His partners, Sir George Horsey** William Hyde Price, *The English Patents of Monopoly* (Clark, NJ: Lawbook Exchange, 2006).

148 **"work for remelting and casting"** Peter W. King, "Sir Clement Clerke and the Adoption of Coal in Metallurgy," *Transactions of the Newcomen Society* 73, no. 1, 2001–2002.

148 **Luckily for Darby** Eugene S. Ferguson, "Metallurgical and Machine-Tool Developments," in Kranzberg and Pursell, eds., *Technology in Western Civilization.*

150 **His greatest contribution to metallurgical history** Cyril Stanley Smith, "Metallurgy in the 17th and 18th Centuries," in Kranzberg and Pursell, eds., *Technology in Western Civilization.*

152 **After nearly ten years of secret experiments** "Benjamin Huntsman" in *Oxford Dictionary of National Biography.*

153 **His furnaces could be made** Smith, "Metallurgy in the 17th and 18th Centuries," in Kranzberg and Pursell, eds., *Technology in Western Civilization.*

153 **He departed from the norm** Joel Mokyr, *The Gifts of Athena: Historical Origins of the Knowledge Economy* (Princeton: Princeton University Press, 2002).

154 **in 1750, when Britain consumed** Pacey, *Maze of Ingenuity.*

154 **"the father of the iron trade"** *The Times*, editorial, July 29, 1856.

154 **a relatively pure form of wrought iron** From Dr. Joseph Gross's description of Wood's process in *Puddling in the Iron Works of Merthyr Tydfil*, quoted at http://www.henrycort.net.

155 **"The puddlers were the artistocracy"** Postan and Habakkuk, *The Cambridge Economic History of Europe* (Cambridge: Cambridge University Press, 1966).

155 **"a peculiar method of preparing"** R. A. Mott and Peter Singer, *Henry Cort, The Great Finer: Creator of Puddled Iron* (London: Metals Society, 1983).

156 **The source of the funds** Newman and Brown, *Britain in the Hanoverian Age.*

156 **Not only had grooved rollers** Jennifer Tann, "Richard Arkwright and Technology," *History: The Journal of the Historical Association* 58, no. 192, February 1973.

156 **"cleansed of sulphurous matter"** R. A. Mott and P. Singer, *Henry Cort, the Great Finer: Creator of Puddled Iron* (London: Metals Society, 1983), quoted at http://www.henrycort.net.

CHAPTER EIGHT: A FIELD THAT IS ENDLESS

159 **Steam engine components were a promising enough source** Raistrick, *Dynasty of Iron Founders.*

160 **by way of comparison, Watt's 1770 salary** Muirhead, *Life of James Watt.*

160 **In 1772, Glasgow's recently established Bank of Ayr** John Lord, *Capital and Steam-Power* (London: Frank Cass, 1966).

160 **Roebuck testified** Scherer, "Invention and Innovation in the Watt-Boulton Steam Engine Venture."

160 **"value the engine at a farthing"** Birmingham Central Library and Adam Matthew Publications, *The Industrial Revolution: A Documentary History. Series One: The Boulton and Watt Archive and the Matthew Boulton Papers from the Birmingham Central Library.*

160 **He persuaded the other claimants** Marshall, *James Watt.*

161 **"The business I am here about"** Ibid.

162 **"we might give up the present patent"** Scherer, "Invention and Innovation in the Watt-Boulton Steam Engine Venture."

162 **"the most important single event"** Eric Robinson, "Matthew Boulton and the Art of Parliamentary Lobbying," *The Historical Journal* 7, no. 2, 1964.

163 **"AND WHEREAS, in order to manufacture"** Scherer, "Invention and Innovation in the Watt-Boulton Steam Engine Venture."

164 **The British Society of Arts** Joel Mokyr, "Mercantilism, The Enlightenment, and the Industrial Revolution," *Conference in Honor of Eli Heckscher,* Stockholm, May 2003.

164 **"a Number of Scientific Gentlemen"** Marshall, *James Watt.*

165 **"I rejoice at the well doing of Willey Engine"** Boulton to Watt, March 1776, in Robinson and Musson, *James Watt and the Steam Revolution.*

165 **"engines of mortality of all descriptions"** Wallace, *Social Context of Innovation.*

167 **"a cylinder attached to a spindle"** Usher, *History of Mechanical Inventions.*

167 **In 1800, boring a 64-inch cylinder** Eugene S. Ferguson, "Metallurgical and Machine-Tool Developments," in Kranzberg and Pursell, eds., *Technology in Western Civilization.*

167 **"unsound, and totally useless"** Robinson and Musson, *James Watt and the Steam Revolution.*

168 **"I wish to do all in the best manner"** Birmingham Central Library and Adam Matthew Publications, *The Industrial Revolution: A Documentary History. Series Three: The Papers of James Watt and His Family Formerly Held at Doldowlod House.*

168 **If the new design caught on** David L. Landes, *The Wealth and Poverty of Nations: Why Some Are So Rich and Some So Poor* (New York: W.W. Norton, 1998).

169 **"With hardly room to move their bodies"** E. D. Clarke, *Tour Through the South of England,* quoted in Howard Jones, *Steam Engines.*

169 **Adventurers, in turn, appointed "captains"** Anthony Burton, *Richard Trevithick, Giant of Steam* (London: Aurum Press, 2000).

169 **"shareholders might grumble"** Ibid.

170 **One of them, the Great County Adit** Ibid.

170 **88 lb. in London** Hills, *Power from Steam.*

171 **"raise as much water as two Horses"** Ibid.

172 **As a result of the hostility** Ibid.

172 **"all the cast iron"** Ibid.

173 **at least one Boulton & Watt engine was too large** Ibid.

173 **It was in response to these demands** Ibid.

173 **"All the world are agape"** Marshall, *James Watt.*

174 **"I think that these mills represent"** Birmingham Central Library and Adam Matthew Publications, *The Industrial Revolution: A Documentary History. Series One: The Boulton and Watt Archive and the Matthew Boulton Papers from the Birmingham Central Library.*

175 **"The technical advance which characterizes"** Lewis Mumford, *Technics and Civilization* (New York: Harcourt, 1934).

176 **"continuous rotary motion"** Lynn White, "The Act of Invention: Causes, Contexts, Continuities, and Consequences," *Technology and Culture* 3, no. 4, March 1962.

176 **Arabs were using them** White, *Medieval Technology and Social Change.*

176 **The earliest visual evidence of a crankshaft** Ibid.

178 **"to render the motion more regular and uniform"** Hills, *Power from Steam.*

178 **Watt had not believed** George Selgin and John Turner, "James Watt as Intellectual Monopolist: Comment on Boldrin and Levine," *International Economic Review* 47, no. 4, November 2006.

179 **"I know the contrivance is my own"** Watt to Boulton, April 28, 1781, in Robinson and Musson, *James Watt and the Steam Revolution.*

179 **"timmer [timber] . . . turned on my little lathey [lathe]"** "William Murdock" in *Oxford Dictionary of National Biography.*

180 **"I wish William could be brought to do"** Ibid.

180 **"the most active man and best engine erector I ever saw"** Birmingham Central Library and Adam Matthew Publications, *The Industrial Revolution: A Documentary History. Series Three: The Papers of James Watt and His Family Formerly Held at Doldowlod House.*

182 **"gave the additional advantage"** Hills, *Power from Steam.*

182 **"for certain new methods"** Ibid.

183 **"I wish you could supply me with a draughtsman"** Samuel Smiles, *Lives of Boulton and Watt. Principally from the original Soho mss. Comprising also a history of the invention and introduction of the steam engine* (Philadelphia: J.B. Lippincott, 1865).

183 **"the neatest drawing I had ever made"** Birmingham Central Library and Adam Matthew Publications, *The Industrial Revolution: A Documentary History. Series One: The Boulton and Watt Archive and the Matthew Boulton Papers from the Birmingham Central Library.*

184 **One consequence was that** Pacey, *Maze of Ingenuity.*

184 **"upon making Engines *cheap,* as well as *good*"** Robinson and Musson, *James Watt and the Steam Revolution.*

185 **"scarce heard in the building where they are erected"** Hills, *Power from Steam.*

186 **"one of the most ingenious"** Birmingham Central Library and Adam Matthew Publications, *The Industrial Revolution: A Documentary History. Series One: The Boulton and Watt Archive and the Matthew Boulton Papers from the Birmingham Central Library.*

186 **"I am more proud"** Birmingham Central Library and Adam Matthew Publications, *The Industrial Revolution: A Documentary History. Series Three: The Papers of James Watt and His Family Formerly Held at Doldowlod House.*

187 **"the millers are determined to be masters of us"** Lord, *Capital and Steam-Power.*

CHAPTER NINE: QUITE SPLENDID WITH A FILE

189 **"The Artist who can make an Instrument"** Jeffrey Kastner, "National Insecurity," *Cabinet Magazine* 22, Summer 2006.

191 **"a LOCK, constructed on a *new* and *infallible* Principle"** J. A. Bramah, *Dissertation on the Construction of Locks* (Goldsmiths-Kress Library of Economic Literature, no. 13077.3, Research Publications: New Haven, CT, 1976).

196 **Even when Bramah promoted Maudslay** Roe, *English and American Tool Builders.*

196 **Unable to persuade Bramah** A. W. Skempton, *Civil Engineers and Engineering in Britain, 1600–1830* (Aldershot, Hampshire, UK, and Brookfield, VT, USA: Variorum, 1996).

199 **"the indefatigable care which he took"** "Henry Maudslay" in *Oxford Dictionary of National Biography. British Academy: from the earliest times to the year 2000.*

199 **"it was a pleasure to see him handle a tool"** "Henry Maudslay" in Ibid.

199 **"a 'critical mass' of inventive activity"** Carolyn C. Cooper, "The Portsmouth System of Manufacture," *Technology and Culture* 25, no. 2, April 1984, quoting William Parker.

200 **A single ship of the line** Ibid.

200 **"Set of engines, tools, instruments"** Ibid.

201 **More significant was Samuel Bentham** John Richards, *A Treatise on the Construction and Operation of Wood-working Machines: Including a History of the Origin and Progress of the Manufacture of Wood-working Machinery* (London, New York: E. & F.N. Spon, 1872).

201 **"a perfect treatise on the subject"** Roe, *English and American Tool Builders.*

201 **In 1786, while in Russia** Ibid.

202 **"*a Propos* of my brother's inventions"** Cooper, "The Portsmouth System of Manufacture."

203 **"roaming on the esplanade of Fort Montgomery"** Richard Beamish and Gerald C. Levy, *Memoir of the Life of Sir Marc Isambard Brunel, Civil Engineer, Vice-President of the Royal Society, Corresponding Member of the Institute of France, &c.* (London: Longman, Green, Longman, and Roberts, 1862).

203 **"two or three at a time"** Cooper, "The Portsmouth System of Manufacture."

204 **The design, however, specified** Ibid.

204 **His machines . . . included power saws** Roe, *English and American Tool Builders.*

204 **Maudslay's fee for constructing the machines** Cooper, "The Portsmouth System of Manufacture."

204 **That agreement guaranteed** "Marc Isambard Brunel" in *Oxford Dictionary of National Biography.*

205 **He didn't come close to recouping** Beamish and Levy, *Memoir of the Life of Sir Marc Isambard Brunel.*

205 **"an output greater"** Roe, *English and American Tool Builders.*

205 **"set fire to the dockyards"** Cooper, "The Portsmouth System of Manufacture."

206 **"new combined steam engines"** Hills, *Power from Steam.*

207 **the planned expansion** "Matthew Murray" in *Oxford Dictionary of National Biography.*

208 **It wasn't until 1788** Selgin and Turner, "James Watt as Intellectual Monopolist: Comment on Boldrin and Levine."

208 **"the ungrateful, idle, insolent Hornblowers"** Birmingham Central Library and Adam Matthew Publications, *The Industrial Revolution: A Documentary History. Series Three: The Papers of James Watt and His Family Formerly Held at Doldowlod House.*

209 **"if patentees are to be regarded"** Stirk, "Intellectual Property and the Role of Manufacturers."

209 **"Our cause is good"** Smiles, *Lives of Boulton and Watt.*

209 **"monstrous stupidity"** Roe, *English and American Tool Builders.*

209 **"I think we should confine our contentions"** Birmingham Central Library and Adam Matthew Publications, *The Industrial Revolution: A Documentary History. Series Three: The Papers of James Watt and His Family Formerly Held at Doldowlod House.*

210 **"a sufficient action against the piston"** Hills, *Power from Steam.*

211 **Maudslay, on the other hand** Skempton, *Civil Engineers and Engineering in Britain, 1600–1830.*

211 **"A zealous promoter of the arts and sciences"** Ibid.

CHAPTER TEN: TO GIVE ENGLAND THE POWER OF COTTON

213 **Lombe was the son of a woolen weaver** Anthony Calladine, "Lombe's Mill: An Exercise in Reconstruction," *Industrial Archaeology* 16, no. 1, Autumn 1993.

214 **A single cocoon of *B. mori*** Yong-woo Lee, *Silk Reeling and Testing Manual,* FAO Agricultural Services Bulletin no. 136, United Nations, Rome, 1999.

214 **Silk from Chinese looms** John Ferguson, "China and Rome" in *Aufstieg und Niedergang der römischen Welt,* vol. 9.2 (Berlin and New York: Walter de Gruyter, 1978).

214 **The Turkish city of Bursa** Robert Sabatino Lopez, "Silk Industry in the Byzantine Empire," *Speculum* XX, January 1945.

214 **In 1665, five Dutch ships** Rudolf P. Matthee, *The Politics of Trade in Safavid Iran: Silk for Silver, 1600–1730* (Cambridge and New York: Cambridge University Press, 1999).

215 **It was Zonca's machine** Usher, *History of Mechanical Inventions.*

215 **"three sorts of engines never before made"** "Thomas Lombe" in *Oxford Dictionary of National Biography.*

216 **The mill, which employed more than two hundred men** Ibid.

216 **"he has not hitherto received the intended benefit"** Smiles, *Men of Invention and Industry.*

216 ***The case of the manufacturers of woolen*** "Thomas Lombe" in *Oxford Dictionary of National Biography.*

217 **Only one silk spinning factory** Abbott Payson Usher, "The Textile Industry, 1750–1830," in Kranzberg and Pursell, eds., *Technology in Western Civilization.*

218 **Even before the Company chose the village of Calcutta** Landes, *Wealth and Poverty of Nations.*

218 **Even then, it made for a very rough weave** Woodruff D. Smith, *Consumption and the Making of Respectability, 1600–1800* (New York: Routledge, 2002).

219 **Between 1700 and 1750** T. Ivan Berend, *An Economic History of Twentieth Century Europe: Economic Regimes from Laissez-Faire to Globalization* (Cambridge: Cambridge University Press, 2006).

219 **The market for cotton** This is a highly abbreviated version of the argument made by the historian Eric Hobsbawm. E. J. Hobsbawm and Chris Wrigley, *Industry and Empire from 1750 to the Present Day* (New York: New Press, 1999).

219 **Those overseas consumers were needed** Angus Maddison, ed., *The World Economy: Historical Statistics* (Paris: Development Centre of the Organisation for Economic Co-operation and Development, 2003). Between 1700 and 1820, British per capita GDP grew by 37%, while the rest of Western Europe grew by less than 19% and the Netherlands *declined* by 14%.

220 **They were the ones who were able to attract the attention** Jan De Vries, "The Industrial Revolution and the Industrious Revolution," *The Journal of Economic History* 54, no. 2, June 1994.

220 **By 2000 BCE** Usher, *History of Mechanical Inventions.*

221 **He never patented** "John Kay" in *Oxford Dictionary of National Biography.*

222 **Inventors in good odor at the Bourbon court** B. Zorina Khan, "An Economic History of Patent Institutions," EH.Net Encyclopedia, March 16, 2008, at http://eh.net/encyclopedia/article/khan.patents.

223 **The significance of this fact for industrialization** Abbott Payson Usher, "The Textile Industry 1750–1830," in Kranzberg and Pursell, eds., *Technology in Western Civilization.* The presumed asymmetry between the productivity of weaving and spinning in eighteenth-century England has recently been questioned and is no longer regarded as unassailable. However, it seems that the weight of the evidence still supports it.

223 **"This is second only to the printing press"** Usher, *History of Mechanical Inventions,* citing Theodore Beck.

223 **The first wheels used to mechanize** Ibid. Lynn White, citing ambiguous illustrations in the windows of the cathedral at Chartres and an earlier regulation in the town of Speyer, gives the date of 1280.

224 **"put [it] between a pair of rollers"** Usher, "The Textile Industry, 1750–1830," in Kranzberg and Pursell, eds., *Technology in Western Civilization.*

225 **In a flash, Hargreaves imagined** "James Hargreaves" in *Oxford Dictionary of National Biography.*

225 **"almost wholly with a pocket knife"** Ibid.

225 **"came to our house and burnt"** Ibid.

225 **"much application and many trials"** Ibid.

226 **"Weavers typically rested and played long"** Landes, *Wealth and Poverty of Nations.*

226 **"When in due course, SAINT MONDAY"** Douglas A. Reid, "The Decline of Saint Monday 1766–1876," *Past and Present* no. 71, 1976.

228 **"I was a barber"** "Richard Arkwright" in *Oxford Dictionary of National Biography.*

228 **As both men later recalled** Ibid.

228 **(Highs's daughter, Jane)** Edward Baines, *History of the Cotton Manufacture in Great Britain* (London: H. Fisher, R. Fisher, 1835).

230 **". . . wee [*sic*] shall not want"** "Richard Arkwright" in *Oxford Dictionary of National Biography.*

231 **He was, partly because of his success with waterpower** Tann, "Richard Arkwright and Technology."

232 **he could scarcely add to or subtract** Ibid.

232 **"no motion can ever act perfectly steady"** Hills, *Power from Steam.*

232 **"earliest steam-powered cotton spinning mill"** Ibid.

232 **somewhere north of £200,000** Tann, "Richard Arkwright and Technology."

233 **"if any man has found out a thing"** "Richard Arkwright" in *Oxford Dictionary of National Biography.*

233 **"There sits the thief!"** R. A. Burchell, *The End of Anglo-America: Historical Essays in the Study of Cultural Divergence* (Manchester, UK, and New York: Manchester University Press and St. Martin's Press, 1991).

233 **"the old Fox is at last caught"** R. S. Fitton, *The Arkwrights: Spinners of Fortune* (Manchester, UK, and New York: Manchester University Press and St. Martin's Press, 1989).

234 **"If yourself or Mr. Watt think as I do"** Ibid.

234 **"Though I do not love Arkwright"** Smiles, *Lives of Boulton and Watt.*

234 **"I have visited Mr. Arkwright"** Birmingham Central Library and Adam Matthew Publications, *The Industrial Revolution: A Documentary History. Series Three: The Papers of James Watt and His Family Formerly Held at Doldowlod House.*

235 **"An engineer's life without patent"** Robinson and Musson, *James Watt and the Steam Revolution.*

235 **"not as the price of a secret"** Birmingham Central Library and Adam Matthew Publications, *The Industrial Revolution: A Documentary History. Series Three: The Papers of James Watt and His Family Formerly Held at Doldowlod House.*

235 **No scientific discovery is ever named** Malcolm Gladwell, "In the Air," *The New Yorker,* May 12, 2008.

236 **"possessed unwearied zeal"** Tann, "Richard Arkwright and Technology."

236 **"A plain, almost gross"** Thomas Carlyle, *Past and Present* (Boston: Houghton Mifflin, 1965).

237 **"a son, brother, or orphan nephew"** Tine Bruland, "Industrial Conflict as a Source of Innovation," in MacKenzie and Wajcman, *The Social Shaping of Technology: how the refrigerator got its hum.*

237 **In the industry's Lancashire heartland** William Lazonick, "The Self-Acting Mule and Social Relations in the Workplace," Ibid.

238 **didn't catch on in Britain** Mokyr, *Lever of Riches.*

238 **"the last of the great inventors"** Usher, "The Textile Industry, 1750–1830," in Kranzberg and Pursell, eds., *Technology in Western Civilization.*

238 **"as soon as Arkwright's patent expired"** Ibid.

239 **In 1551 Parliament passed legislation** Mokyr, *Lever of Riches.*

240 **Not only was Richard Hargreaves's original spinning jenny destroyed** Jeff Horn, "Machine-breaking in England and France During the Age of Revolution," *Labour/Travail* 55, Spring 2005.

240 **Normandy in particular** Ibid.

240 **"the machines used in cotton-spinning"** Ibid.

240 **"he had favored machines"** Ibid.

240 **"prejudice against machinery"** Ibid.

240 **"collective bargaining by riot"** Kevin Binfield, *Writings of the Luddites* (Baltimore and London: Johns Hopkins University Press, 2004).

241 **Handloom weavers had been earning** Sale, *Rebels Against the Future.*

241 **"mee-mawing"** Ibid.

241 **more than half of all the land then in cultivation in England** Ibid.

242 **The lack of a patent** Usher, *History of Mechanical Inventions.*

242 **In the late 1770s, they petitioned Parliament** Binfield, *Writings of the Luddites.*

243 **The stockingers began in the town of Arnold** Ibid.

243 **The attacks continued throughout the spring** Horn, "Machine-breaking in England and France During the Age of Revolution."

243 **That November, a commander** Ibid.

243 **"2000 men, many of them armed"** Binfield, *Writings of the Luddites.*

243 **Manchester was further down the path** Ibid.

243 **Manchester alone had more than three thousand men** Sale, *Rebels Against the Future.*

244 **In January, the West Riding of Yorkshire** Binfield, *Writings of the Luddites.*

244 **"Whereas by the charter"** A. Aspinall and E. Anthony Smith, eds., *English Historical Documents XI, 1783–1832* (New York: Oxford University Press, 1959).

245 **This made them not only self-interested** Horn, "Machine-breaking in England and France During the Age of Revolution."

245 **"I do not mean to say, that parties of Luddites"** Aspinall and Smith, eds., *English Historical Documents XI, 1783–1832.*

246 **"You must raise your right hand"** Ibid.

247 **In 1813, there were 2,400 power looms** Usher, "The Textile Industry 1750–1830," in Kranzberg and Pursell, eds., *Technology in Western Civilization.*

247 **"a steam-loom weaver"** Hills, *Power from Steam,* quoting Baines's 1835 *History of the Cotton Manufacture in Great Britain.*

247 **During the century and a half** Clark, *Farewell to Alms.*

CHAPTER ELEVEN: WEALTH OF NATIONS

250 **nothing about the forging of iron** David Warsh, *Knowledge and the Wealth of Nations: A Story of Economic Discovery* (New York: W. W. Norton, 2006).

251 **David Ricardo predicted** Clark, *Farewell to Alms.*

252 **The second component, growth in capital** Warsh, *Knowledge and the Wealth of Nations.*

252 **Solow first assumed** Ibid.

254 **"the mass of persons with intermediate skills"** Hobsbawm and Wrigley, *Industry and Empire: from 1750 to the Present Day.*

254 **preindustrial Britain exhibited a fair bit** F. F. Mendels, "Social mobility and phases of industrialization," *Journal of Interdisciplinary History* 7, 1976.

254 **"craftsman's sons became laborers"** Clark, *Farewell to Alms.*

255 **A recent World Bank analysis** Kirk Hamilton, et al., *Where Is the Wealth of Nations? Measuring Capital for the XXI Century* (Washington, D.C.: World Bank, 2005).

255 **India was home, in 1700** Maddison, ed., *The World Economy: Historical Statistics.*

256 **Solow's fundamental growth equation** Warsh, *Knowledge and the Wealth of Nations.*

257 **Kremer's model made two assumptions** Kremer, "Population Growth and Technological Change."

259 **It's not as if Kremer was unaware** Kremer, "Population Growth and Technological Change."

259 **But China had, and has, huge coal deposits** Kenneth Pomeranz, *The Great Divergence: Europe, China, and the Making of the Modern World Economy* (Princeton: Princeton University Press, 2000).

260 **And even though the barbarian invasions** Ibid.

260 **Needham's conclusion** Joseph Needham, "The Pre-Natal History of the Steam Engine," *Newcomen Society Transactions* 35, no. 49, 1962–63.

260 **By 400 CE they had developed a system of water "levers"** Mokyr, *Lever of Riches.*

261 **"let it [the box bellows] be furnished"** Ian Inkster, "Indisputable Fea-

tures and Nebulous Contexts: The Steam Engine as a Global Inquisition," *History of Technology* 25, 2004.

261 **"The Chinese had already recognized"** Pomeranz, *Great Divergence.*

261 **The Chinese could have a bellows** Kent G. Deng, "Why the Chinese Failed to Develop a Steam Engine," *History of Technology* 25, 2004.

262 **China's master artisans were so severely handicapped by illiteracy** Clark, *Farewell to Alms.*

262 **The draw bar was not a complicated device** Mokyr, *Lever of Riches.*

262 **"The absence of political competition"** Joseph Needham, Guohao Li, Meng-wen Chang, Tienchin Ts'ao, and Tao-ching Hu, *Explorations in the History of Science and Technology in China: A Special Number of the "Collections of Essays on Chinese Literature and History"* (Shanghai: Shanghai Chinese Classics, 1982).

263 **Europe's fragmented system of sovereign states** E. L. Jones, *The European Miracle: Environments, Economies, and Geopolitics in the History of Europe and Asia* (Cambridge and New York: Cambridge University Press, 2003).

263 **Bertrand Russell translated the Chinese term** Mokyr, *Lever of Riches.*

264 **From 1600 to 1650, the Dutch government** Geoffrey Parker, *Europe in Crisis, 1598–1648* (Oxford and Malden, MA: Blackwell, 2001).

264 **Petra Moser, now a professor** Petra Moser, "How Do Patent Laws Influence Innovation? Evidence from Nineteenth Century World's Fairs," *The American Economic Review* 95, no. 4, September 2005.

264 **"would have been silly"** Teresa Riordan, "Patents: An Economist Strolls Through History and Turns Patent Theory Upside Down," *New York Times,* September 29, 2003.

265 **From there on, Britain took over the lead** Kenneth Romer, "Increasing Returns and Long-Run Growth," *Journal of Political Economy* 94, no. 5, October 1986.

265 **The most intriguing candidate** N.F.R. Crafts, "Macroinventions, Economic Growth, and 'Industrial Revolution' in Britain and France," *Economic History Review* 58, no. 3, 1995. One estimate has the Netherlands with GDP per capita of $2,130 in 1700 and $1,838 in 1820, expressed in 1990 U.S. dollars.

266 **In 1789, the year of the Revolution** Melvin Kranzberg, "Prerequisites for Industrialization," in Kranzberg and Pursell, eds., *Technology in Western Civilization.*

266 **By the same year, however** Crafts, "Macroinventions, Economic Growth, and 'Industrial Revolution' in Britain and France."

266 **Thus, in part because of lower interest rates** Kranzberg, "Prerequisites for Industrialization."

266 **Watt was simultaneously a brilliant engineer** Pacey, *Maze of Ingenuity.*

267 **Among other things, the project provided Diderot** E. S. Ferguson, "The Mind's Eye: Nonverbal Thought in Technology."

267 **"to offer craftsmen the chance to learn"** Bertrand Gille, *The History of Techniques* (New York: Gordon and Breach Science Publishers, 1986).

267 **Between 1740 and 1780** Mokyr, *Lever of Riches.*
268 **the French did not lionize their inventors** Ibid.
268 **"to deprive England of her steam engines"** Carnot, quoted in Inkster, "Indisputable Features and Nebulous Contexts: The Steam Engine as a Global Inquisition."
268 **"every novel idea"** Fritz Machlup, "The Patent Controversy in the Nineteenth Century," *Journal of Economic History* 10, no. 1, May 1950.
269 **From 1793 to 1800, in fact** Khan, "An Economic History of Patent Institutions."
269 **"When the revolutionary and Napoleonic wars ended"** Jeff Horn, *The Path Not Taken: French Industrialization in the Age of Revolution, 1750–1830* (Cambridge, MA: MIT Press, 2006).
270 **"the Republic does not need savants"** Mokyr, "The Great Synergy." The phrase supposedly originated at Lavoisier's trial, and it should be noted that the Académie would be reconstituted, under a different name.

CHAPTER TWELVE: STRONG STEAM

272 **"I am exceedingly shocked"** Birmingham Central Library and Adam Matthew Publications, *The Industrial Revolution: A Documentary History. Series One: The Boulton and Watt Archive and the Matthew Boulton Papers from the Birmingham Central Library.*
273 **There he found himself, on behalf of the Prince-Elector** Lienhard, *How Invention Begins.*
274 **Caloric theory held that heat was latent** Hills, "The Development of the Steam Engine from Watt to Stephenson," *History of Technology* 25, 2004.
275 **No matter how many times engineers observed** Ibid.
276 **"the elastic force of steam"** Hills, *Power from Steam.*
276 **Matthew Boulton himself intercepted Murdock** Eugene S. Ferguson, "Steam Transportation," in Kranzberg and Pursell, eds., *Technology in Western Civilization.*
277 **"to grant patents for useful inventions"** Kenneth W. Dobyns, *The Patent Office Pony: A History of the Early Patent Office* (Fredericksburg, VA: Sergeant Kirkland's Museum and Historical Society, 1994).
277 **"Each revolution of the axle tree"** Thompson Westcott, *Life of John Fitch, the Inventor of the Steam-boat* (Philadelphia: J.B. Lippincott, 1857).
278 **"the exhibition yesterday"** Ibid.
278 **"provide limited patents"** Ibid.
280 **"If nature has made any one thing"** S. E. Forman, *The Life and Writings of Thomas Jefferson, Including All of His Important Utterances on Public Questions, Compiled from State Papers and from His Private Correspondence* (Indianapolis: The Bobbs-Merrill Company, 1900).
282 **In 1792, the official cost of a patent** Christine MacLeod, et al., "Evaluating Inventive Activity: The Cost of Nineteenth-Century UK Patents and

the Fallibility of Renewal Data," *Economic History Review* LVI, no. 3, Aug. 2003.

282 **The cost of a U.S. patent application** Khan, "An Economic History of Patent Institutions."

283 **"I have in my bed viewed the whole operation"** Ferguson, "The Mind's Eye: Nonverbal Thought in Technology."

284 **"If the bringing together under the same roof"** Carroll W. Pursell, *Technology in America: A History of Individuals and Ideas* (Cambridge, MA: MIT Press, 1981). It should be noted that Jefferson's complaint was as personal as it was political; in 1806, he had built his own mill, using some of the same methods as Evans, who sent the then sitting president a bill for licensing his patented technology.

284 **His goal had been to earn his living** "Oliver Evans" in John A. Garraty, Mark C. Carnes, and American Council of Learned Societies, *American National Biography* (New York: Oxford University Press, 1999).

285 **"Boiler with the furnace in the center"** Greville Bathe and Dorothy Bathe, *Oliver Evans: A Chronicle of Early American Engineering* (Philadelphia: The Historical Society of Pennsylvania, 1935).

285 **"the more the steam is confined"** Hills, *Power from Steam*, citing *The Emporium of Arts and Sciences* 4, 1813.

286 **the first significant improvement over Watt's linkage** Eugene S. Ferguson, "Kinematics of Mechanisms from the Time of Watt," *United States National Museum Bulletin* 288, Paper 27, 1962.

286 **"in poverty at the age of 50"** Bathe and Bathe, *Oliver Evans*.

287 **"And it shall come to pass"** Ibid.

287 **"the time will come"** Ibid.

288 **"In 1794–95, I sent drawings"** Ibid.

289 **"disobedient, slow, obstinate"** Anthony Burton, *Richard Trevithick*.

289 **Gilbert explained that with each stroke** Kerker, "Science and the Steam Engine."

291 **"Captain Dick got up steam"** Anthony Burton, *Richard Trevithick*.

292 **Two of Trevithick's drivers** Hills, "The Development of the Steam Engine from Watt to Stephenson."

292 **One of the more avid users was the Coalbrookdale foundry** "Richard Trevithick" in *Oxford Dictionary of National Biography*.

292 **"Mr. B. & Watt"** Hills, *Power from Steam*.

294 **Also, the grade was extremely gentle** Anthony Burton, *Richard Trevithick*.

294 **"makes the draft much stronger"** L.T.C. Rolt, *George and Robert Stephenson: The Railway Revolution* (London: Longmans, 1960).

295 **"yesterday we proceeded"** National Museum of Wales, "Richard Trevithick's Steam Locomotive," 2008, online article at http://www.museumwales.ac.uk/en/rhagor/article/trevithic_loco/.

296 **"My predecessors . . . put their boilers in the fire"** Selgin and Turner, "James Watt as Intellectual Monopolist: Comment on Boldrin and Levine."

296 **By 1812, Trevithick was boasting** Hills, *Power from Steam.*

297 **Trevithick was forced to flee north** "Richard Trevithick" in *Oxford Dictionary of National Biography.*

298 **"half-drowned, half-dead and the rest devoured by alligators."** Francis Trevithick, *Life of Richard Trevithick, with an Account of His Inventions* (London and New York: E. & F.N. Spon, 1872).

299 **$100,000 in current dollars** This is calculated as the increase in the index of average earnings, rather than the retail price index. The actual figure is approximately £66,000 a year. www.measuringworth.com/ppoweruk

301 **Slightly less eyebrow-raising** Rolt, *George and Robert Stephenson.*

302 **The first common carrier to realize this** Eugene S. Ferguson, "Steam Transportation," in Kranzberg and Pursell, eds., *Technology in Western Civilization.*

304 **"From professors of philosophy"** Quoted in Rolt, *George and Robert Stephenson.*

305 **"Rely upon it, locomotives"** Archives of the Science Museum, London, at http://www.makingthemodernworld.org.uk~/01.st.04

309 **"the water flying in all directions"** Samuel Smiles, *The Life of George Stephenson, Railway Engineer* (London: J. Murray, 1857).

310 **"he had the lives of many men in him"** Lawrence F. Abbott, *Twelve Great Modernists: Herodotus, St. Francis, Erasmus, Voltaire, Thomas Jefferson, John Marshall, François Millet, George Stephenson, Beethoven, Emerson, Darwin, Pasteur* (New York: Doubleday, Page, 1927).

EPILOGUE: THE FUEL OF INTEREST

313 **This tempted John Ericsson** Lynwood Bryant, "The Role of Thermodynamics in the Evolution of Heat Engines," *Technology and Culture* 14, no. 2, April 1973.

313 **"if not absolutely perfect in its action"** Hills, *Power from Steam.*

314 **The ten-thousand-horsepower monster** Ibid.

317 **"Small changes in the parameters of the economy"** Paul Krugman, "Increasing Returns and Economic Geography," *The Journal of Political Economy* 99, no. 3, June 1991.

318 **This is an argument starter** Deepak Lal, "In Defense of Empires," in A. J. Bacevich, ed., *The Imperial Tense* (Chicago: Ivan R. Dee, 2003).

318 **"no nation has been very creative"** D.S.L. Cardwell, *Turning Points in Western Technology: A Study of Technology, Science and History* (New York: Science History Publications, 1972).

318 **In 1700, when Great Britain's per capita** All following statistics are drawn from Maddison, ed., *The World Economy: Historical Statistics.*

319 **The age of scientific invention** Mokyr, *Lever of Riches.*

319 **In 1850, France alone issued** Khan, "An Economic History of Patent Institutions," EH.Net Encyclopedia, edited by Robert Whaples, March 16, 2008, at http://eh.net/encyclopedia/article/khan.patents.

319 **"regard discovery and invention"** Ibid.

321 **"It is not the man of the greatest natural vigour"** Quoted in Katrina Honeyman, *Origins of Enterprise: Business Leadership in the Industrial Revolution* (Manchester: Manchester University Press, 1982).

323 **"Man . . . is not the only animal who labors"** Lincoln, Basler and Abraham Lincoln Association (Springfield, IL), *Collected Works* (New Brunswick, NJ: Rutgers University Press, 1953), vol. 3.

INDEX

ABOUT THE AUTHOR

WILLIAM ROSEN, the author of the award-winning history *Justinian's Flea: Plague, Empire, and the Birth of Europe,* was an editor and publisher at Macmillan, Simon & Schuster, and the Free Press for nearly twenty-five years. He lives in Princeton, New Jersey, and can be visited at www.mostpowerfulidea.com.